D0919843

The Human Side of Just-In-Time

THE HUMAN SIDE OF JUST-IN-TIME

HOW TO MAKE THE TECHNIQUES REALLY WORK

Charlene B. Adair-Heeley

amacom
American Management Association

This book is available at a special discount when ordered in bulk quantities. For information, contact Special Sales Department, AMACOM, a division of American Management Association, 135 West 50th Street, New York, NY 10020.

This publication is designed to provide accurate and authoritative information in regard to the subject matter covered. It is sold with the understanding that the publisher is not engaged in rendering legal, accounting, or other professional service. If legal advice or other expert assistance is required, the services of a competent professional person should be sought.

Library of Congress Cataloging-in-Publication Data

Adair-Heeley, Charlene.
The human side of just-in-time : how to make the techniques really work / Charlene Adair-Heeley.
p. cm.
Includes index.
ISBN 0-8144-5031-8
1. Just-in-time systems. I. Title.
TS157.A27 1991 90-56200
658.5′6—dc20 CIP

Printing number

10 9 8 7 6 5 4 3 2 1

To **Dick**
whose love, support, and
patience made this book possible.

To **Ronda**
whose strength,
bright intellect and creativity,
and love give me inspiration.

To **Daren**
whose balance of life,
appreciation of nature and music,
and love remind me to "stop and smell the roses."

To **Charles and Eloise Blass,**
whose love, values, and belief in me
gave me the ability to believe in myself.

Contents

Foreword

We think of ourselves as managers of the 1990s. We have heard of Just-In-Time. We know it is being taught in business schools. We have read about it in business magazines and have heard it mentioned at seminars. Others seem to have been achieving surprising benefits by adopting Just-In-Time methodologies in their companies.

But, for many of us, Just-In-Time remains a vague concept. Traditional manufacturing line supervision is obsolete. Psychologically comforting piles of raw material inventory and half-finished assemblies are eliminated. Production employees spend time in meetings problem solving. Lines can be shut down without hours of top management review and consideration. Production engineering must respond to line problems immediately.

This sometimes results in negative reactions: Management authority will be lost! Costs will go up! Productivity will be lost! Chaos will reign!

I faced just such reactions in attempting to introduce JIT to my company in 1987. Truth be told, some of the reservations were mine. But I was committed to bringing the benefits of JIT to my company and was confident that we had the ability to succeed—but not without an internal champion supported by a practical, down-to-earth, hands-on consultant.

We retained Charlene Adair-Heeley, who came to us with the clear and concise vision of the how-to of JIT presented in this book—addressing not only the technical aspects of JIT but also the how-to associated with changing the culture and getting the commitment to make it work. She demonstrated, as she does in this book, that Just-In-Time is a business management philosophy that fits perfectly with our overall goals and commitment to product integrity, quality, customer service, operating efficiency, employee involvement, and vendor participation.

The results were impressive. From a six-month trial project on one assembly line, our program grew over a two-and-one-half-year period to encompass all manufacturing and warehousing operations, materials management, significant vendor involvement, and even customer service and other administrative functions. In a company employing several hundred production employees in both high-volume low technology and low-volume high technology operations and sustaining a 20 percent annual growth rate, we were able to:

- Reduce floor space requirements up to 10 percent.
- Virtually eliminate work-in-process inventories.
- Reduce raw material inventories.
- Improve order fulfillment levels.
- Achieve zero-defect production on lines for as long as eleven consecutive months.
- Improve overall quality levels as evidenced by customer complaint ratios per 1,000 units sold.
- Increase productivity (measured in labor hours/unit) by up to 10 percent.
- Reduce line-down problems, and length of down time and rework.
- Improve introduction of new products to manufacturing.
- Improve employee participation and job satisfaction.
- Certify key vendors to ship to production floor, eliminating costs and delay.

These results are all documented against baseline data, and percentage improvements are very significant. Total cost savings exceeded any other program in that time period and represented annual savings of seven times the investment.

At the end of our two-and-one-half-year implementation program, JIT was a routine method of doing business in our company. We were able to disband special implementation programs.

The manager or executive who feels it cannot be done in his company, or who hesitates to take the first step because of a lack of full understanding of JIT *must* read this book. Charlene Adair-Heeley's consulting record proves it can be done in *any* business, of *any* size, and in *any* industry, including service industries. Her book helps you understand how. In simple, concise language without the aura or mystique often associated with such material, Charlene gives you the background to make informed decisions on JIT implementation. Her text will bring together the bits and pieces of JIT knowledge you may already have and help you understand the entire concept of JIT philosophies and methodologies and how JIT fits with your other business systems and your management goals.

Reading this book could help you make a decision to embark on a JIT implementation program that will be invaluable to your company in remaining competitive, serving your customers, and providing a stimulating and satisfying work environment for your employees. It *can* be done in your company as it was in mine.

Bruce A. Murray
Senior Vice President
Business Manager—Surgical Products
Valleylab, Inc.

Preface

The Human Side of Just-In-Time is written for the action-oriented manager in today's business world. For the first time, the process for achieving long-lasting success with Just-In-Time (JIT) through people is described in a step-by-step fashion, utilizing case histories drawn from my experience. Just-In-Time promotes a culture of continuous improvement through 100 percent involvement of 100 percent of your people in identifying and solving problems as they make proactive improvements. A war on waste is launched, and 100 percent quality is the goal.

Change is inevitable if companies are to survive amid fierce competition. The market requirements for higher quality, better delivery, more flexibility, and lower costs can be met with Just-In-Time principles and techniques. In this book, you will learn how to tap that often untapped resource—your people—as the *only* way to achieve long-term competitive excellence with JIT. Simply implementing a new set of manufacturing techniques will produce only short-term success; for long-term success you must change the way work is done and the culture of the organization so that everyone is committed to continually improving themselves, eliminating wasteful activities, and achieving 100 percent quality 100 percent of the time.

As you apply the tools and follow the process of changing the culture that is described in this book, you will observe astounding, measurable improvements in your company. You will find that these tools and principles are very effective in nonmanufacturing areas and in service companies; they do not apply just to manufacturing businesses.

The implementation process and tools described in this book will help you answer the following questions:

1. How should you begin a JIT implementation?
2. How does JIT work in your industry?
3. How do you select the issues to work on and determine priorities?
4. Who should be involved?
5. When is the company ready?

6. Where within the company should you begin?
7. What is the cost/payback?
8. How do you maintain momentum after initial JIT implementation?

This approach for implementing JIT is tried and proved. At each step, people will be the focus. Involvement by people at every level of the organization has been described as an attribute; this book describes how to utilize involvement as a tool with which to implement the culture and get the commitment to make the techniques of Just-In-Time *really* work. Your people are the key to unlocking the door for future success and, possibly, survival; so read on and act quickly. Good luck as you become a catalyst for change in your organization. It takes perseverance and a lot of hard work. Yet nothing can be more rewarding than seeing the results of your efforts. I know. I have been there.

Charlene B. Adair-Heeley
Nahant, Massachusetts

Acknowledgments

As this book approaches completion, the time has come to reflect on the many special people who have contributed to my career experiences (on which this book is based) and to the production of this book.

A special thanks to Jon Zonderman, who patiently edited the book and assisted greatly in organizing my thoughts and transferring them to paper. The Rath & Strong office staff should be commended for the painstaking effort of getting those words and pictures on paper so professionally.

The Rath & Strong marketing staff and my consulting colleagues provided support, assistance, and learning beyond expectation.

Many clients contributed significantly as we worked together to create new solutions for their companies. It would be impossible to name them all, but some of the key contributors to the material in this book include Ed Costa and Dave Knight of AGFA Compugraphic; Bruce Murray and Dave Kuhn of Valleylab, Inc.; Charles Folkman of Pfizer Hospital Products Group; Len Stuessel of General Dynamics; Wayne Fortune of Hutchison Technology; and Wally Pattyson of Thomas & Betts.

Key turning points and opportunities to learn and grow were made possible by the Oliver Wight Associates, Ken McGuire, Dave Buker, and Hank Greenfield, Warren Darress, and Jeff Duerr of Hewlett Packard. Special thanks to Don Curtis of Hewlett Packard, a strong leader who really cares about people and to Randy Heffner, a brilliant visionary, formerly with Hewlett Packard and now with NEXT.

I thank all of you who played a part in creating the "script," especially all of the team members in many companies who really are the ones who made things happen.

I look forward to continued success for each of you as we all continue to improve and succeed in today's competitive world.

Chapter 1

How to Get and Keep a Competitive Edge With JIT

Recently, my friend Charlie, a manager in a manufacturing company, traveled to visit a company that supplies crucial parts for his company's product. He was particularly looking forward to this trip to the midwestern town because he had friends from a previous job who worked at XYZ Manufacturing Company there—in fact, he wondered why he had not heard from them in a while and decided to allow time to stop by for a visit. As Charlie drove into town, he was distraught to find that stores were for rent and needed paint. Many homes were deserted and in disrepair, the FOR SALE signs long since gone. The large plant where his friends had worked, once the lifeblood of the community, sat idle.

Charlie stopped at a local restaurant to get a bite to eat. One of his friends, Joe, from XYZ was sitting at the bar. He looked so exhausted and worn-out that Charlie hardly recognized him!

They began to talk. Joe told Charlie that he was the last of the old group left in town; he owned a house and would have to walk away from it when he found another job. Joe wondered how his former employer could have gone from being profitable and thriving to nonexistent in only two years.

He began to have flashbacks to the good old days at XYZ. Management was very experienced and made all decisions and directed all work. Nobody questioned authority. In fact, anyone who did was considered a troublemaker. Employees who suggested an idea for improvement were told, "I'll get back to you." Of course, management never did. After all, who were these people to think they could solve management and engineering problems?

Charlie began to feel nervous.

Employees at XYZ were expected to work as fast as possible and were measured on efficiencies—the number of things built during a certain time period compared to a standard set by engineers. People had learned to come to work, work fast, and follow the rules.

Managers were evaluated mainly on their fire-fighting ability. The ones who could "make the impossible happen" got promoted. Shipments were the key—selected shipments in some cases to make monthly goals. End-of-the-month push happened every month.

Charlie thought about his promotions at XYZ, achieved through extremely good fire fighting. In fact, this supplier visit was to fight a fire; it had been planned at the last minute to solve a problem with the supplier and once again save the day!

XYZ inspectors were responsible for quality; they were expected to catch and repair problems, guaranteeing 100 percent quality to the customer. Over the years, many checks, tests, and double-checks had been added to prevent defective products from slipping through. Controls were in place in the form of elaborate computer systems, paperwork, and fences around inventory. More controls were added as the existing ones failed to solve problems in the manufacturing process and customer service.

The pit in Charlie's throat was getting larger. He thought to himself, "In my present company, cycle times and costs have been going up. I wonder if some of our checks to catch problems are unnecessary."

Success factors were internally focused at XYZ—no one had the time to look at the "big picture." Proactive thinking was rare, and so was change.

Joe began to look very sad as he explained to Charlie that XYZ's market share had begun to dwindle—costs went up and quality went down. Management laid off people to reduce costs, but the business did not improve. More cutbacks and layoffs occurred. The competition had passed the company by before management realized what was happening.

The company president of XYZ read about Just-In-Time but decided that this change process didn't apply to this company, which had low volume and built product to order. Suppliers couldn't possibly deliver Just-In-Time—none of them were even close by—and zero inventories certainly didn't make sense—parts shortages were already a problem.

XYZ management just could not imagine managing in a different way; the present style had worked for thirty-five years! The idea of giving up some control and letting employees get involved seemed foolish and too risky.

As time went on, customer service at XYZ continued to decrease while the competition was providing better-quality product at lower prices with good deliveries. The company didn't make it, and Joe was in charge of closing the plant.

By this time, Charlie's heart was beating at a fast pace. Joe could have been describing *his* present company!

How JIT Can Help You Beat the Competition

You may have found yourself in a situation similar to Joe's. However, top management in your company has seen the light and recognized that things must change if the company is to remain competitive.

Just-In-Time concepts and techniques provide a way not only to survive but to beat the competition by creating a culture of continual improvement, involving 100 percent of the people, eliminating waste, and reaching for 100 percent quality. The key is to recognize the need for change early—before it is too late. If you are ready for the challenge, this book will show you how to begin this critical change process.

I am going to make a few assumptions here.

1. Top management of your company has heard about JIT and has been open-minded enough not to be like the top management of the company I just described. Top management has said, "There seems to be some value in this JIT idea. We believe it can work here. Let's make it happen."
2. You are involved in making it happen.
3. Like most people in your position, you really do not know where to begin with such a challenge. Possibly, through reading and seminars, you know that not only is Just-In-Time a set of tools and techniques for improving quality and manufacturing flow and for reducing inventory but that there is a heavy "people" component to it as well.

You may feel unprepared, scared, and somewhat alone. After all, people issues are the domain of the human resources manager. You are not trained in organizational development; you are trained to make things. You may feel a little like a fish out of water with some of the things I am about to say.

I'm reminded of the high school science teacher who recently exclaimed, "I was trained to teach science—a subject I love—to high school students. Now I find myself spending time dealing with teenage pregnancy and drug problems. No one trained me to be a social worker, nor am I sure that I want to be one!"

Even when Just-In-Time gets into what you are most comfortable with—what I call the tools and techniques of flow—you may not be sure you understand or agree with it. Many of the things I suggest as necessary for Just-In-Time to happen go against everything you have ever learned about manufacturing.

All I can say at this point is, it works. Companies I have worked with or know about personally have had improvements in market share and manufacturing quality and reductions in cycle times and cost similar to these:

- 100 percent quality improvement
- 83 percent reduction in cycle time
- 40 percent reduction in cost
- 20 percent increase in market share

Because Just-In-Time is such a radical change from what you have known how to do—and have done very well—throughout your career as a manager in a manufacturing business, this book has to give you more than just forms, checklists, and helpful hints to get small improvements. Instead, it must provide the opportunity to explore how things are done and even why they are done before the benefits of doing them in particular ways becomes apparent.

The examples I use in this book come from three companies: Hewlett-Packard, where I worked for a number of years; AGFA Compugraphic Corp., a client; and Valleylab, also a client. These three companies illustrate the process of implementing JIT, and I repeatedly use examples from their experiences, rather than just isolated examples of success or failure, so you will have a feeling for how the process works.

JIT is a state of mind, a philosophy, that must be created within a company in order for the particular tools and techniques to work. That job does not belong solely to the human resources department, the traditional "people" people. And it does not belong only to top management, who must "walk it as well as talk it." That job belongs to you, because you are the leader working with each person, each "expert" who knows more about his or her job than anyone else in the company. You have to make all those experts function as a team. You have to make them all understand that being the best they can be as individuals means being the best they can be as one person working in harmony with others.

These experts exist throughout the company. JIT is not just for manufacturing; it applies to every area of the business (Chapter 10 describes how JIT works in nonmanufacturing areas of the company, as well as in nonmanufacturing, service organizations).

In this book, I describe some specific steps for addressing the issue of changing the state of mind of an organization. You will discover how recognizing the people issues in your company can result in specific improvements that will show up on the balance sheet and in your market arena.

Now it is important to define this much-misunderstood concept of Just-In-Time.

Just-In-Time
<u>is</u>
a STATE OF MIND
<u>for</u>
achieving competitive excellence
<u>by</u>
creating an attitude of continuous improvement
through 100 percent involvement
<u>to</u>
eliminate all waste
institutionalizing only value-adding activities
<u>with</u>
100 percent quality—NOTHING LESS!

Tying People Into JIT to Make the Techniques Work

Just-In-Time is a state of mind that exists within an organization—not just a set of techniques. If the new state of mind is achieved and the organization's culture is changing, mastering the techniques will be easy. Most JIT techniques make sense and can be learned and implemented readily. If the desired state of mind is achieved and all employees feel confident about the reasons for JIT and feel a sense of ownership in JIT, changes can become permanent. But if an organization concentrates only on techniques, changes implemented will last only as long as a consultant or manager is driving them. When this leader leaves, so will the ownership and commitment.

Using JIT to Stay Ahead of the Competition

Why should your company embark on such an all-encompassing effort? The biggest reason is competition that may be taking market share. Sometimes the competition speeds by before a company even realizes that it is in trouble. When this happens, many companies cannot change quickly enough to survive. Some do survive; Harley-Davidson is an excellent example.

From Riches to Rags to Riches Again

For many years, Harley-Davidson was synonymous with motorcycling and carried an assurance of quality, durability, and value. Harley controlled 70 percent of the market in the mid-1950s. However, like many other American companies, it became complacent; from the late 1950s through the 1970s, Harley failed to react to economic and social changes developing throughout the world.

By 1981, Harley-Davidson's share of the U.S. motorcycling market had fallen to only 5.2 percent. Some of the problems experienced at that time included use of only a fraction of the employee assets, parts shortages, large inventories of some parts and critical shortages of others, high rework and scrap, and a large inspection department that had the responsibility but not the means for controlling quality throughout the shop. Harley became associated with excessive engine vibration and breakdowns, and the name lost its luster.

By the early 1980s, Harley was producing overpriced, low-quality products, while Japan was doing just the opposite. In 1982, 1,600 of the 3,800 employees were laid off. The years 1981 and 1982 saw staggering losses, and the survival of the company was questionable.

The direction of the company was changed in 1982. Harley-Davidson embarked on a major quality-improvement program that focused on employee involvement, Just-In-Time, and Statistical Process Control. By focusing on quality, improvements and cost reductions came naturally.

Harley convinced President Reagan to sign a five-year import tariff on large motorcycles to take effect in 1983; this gave the company some time to make the improvements necessary to be competitive once again.

Management decided to make no change unless people affected by that change were involved in identifying and solving problems while making proactive improvements. More communication of information to all employees became the standard. Harley attributes improvements such as reduced carrying costs, improved cash flow, increased inventory turns, reduced scrap and rework, and reduced warehouse requirements to this employee involvement. The most outstanding benefit by far was increased quality: Defects were reduced by 53 percent and warranty claims by 36 percent by August 1985.

Harley used a concept known as manufacturing centers to allow employees to control parts from raw material to finished product. Employees were trusted to manage these centers, creating a great deal of pride and ownership.

Continuous improvement is a way of life at Harley-Davidson today. The year 1986 was profitable, and in 1987 the company asked the government to release the tariff—early! It was ready to compete.

I believe Harley-Davidson is alive and well due to the implementation of Just-In-Time concepts and techniques. The three lessons to be learned here are:

1. Never discount the competition.
2. Always watch the competition.
3. Be willing to change. Companies that hold on to the theory that "this is the way we've always done things—it should still work" are not surviving in today's competitive markets.

Benchmarking

Xerox has instituted a practice of "benchmarking" as a way of setting new standards for excellence. Benchmarking involves finding the best competitor in each area of the business and making that the standard to strive for. Xerox's use of benchmarks extends far beyond copiers. L. L. Bean, the catalogue clothing company, is the benchmark for distribution operations. American Express is the benchmark for billing. Xerox even benchmarks legal and public relations.

Encouraging Continuous Improvement for Long-Term Success

I travel to many different manufacturing facilities. On occasion, I enter a plant that is really outstanding—housekeeping is impeccable, the people are happy and ready to talk (even to a consultant in a suit!), inventories are organized and in control. By watching, I can determine the flow of product in this area. I am very impressed.

I ask, "How have you accomplished so much? What is the secret to your success?" The reply is often something like, "Next week we're getting a new tool that we designed and engineering ordered for us"; "We plan to try moving that workstation a little closer over there"; or "Our team is attending some additional training next week." I can't get them to talk about how they got where they are—they are focused on where they are going. I leave this plant very excited.

The next day, I visit another plant. Here the housekeeping is not very good—aisles are narrow, and boxes and inventory are in the way. The place is not clean. One pallet of inventory has a considerable amount of dust on it. The people look down when they see me coming. You can almost see them thinking, "What's management up to now, bringing in this consultant?" Everyone is working frantically, but I cannot tell anything about the product flow. Finally, I find someone who is willing to talk to me. When I begin to ask the questions, "Why hasn't this inventory been used? Why isn't the housekeeping better? Why isn't the product flow better defined?" the answers are all the same: "We can't do that because . . ." I call these the "because plants."

What is *really* different about these two facilities is the state of mind.

The first group is always looking for improvements; the second is collecting excuses.

A basic premise upon which the successful JIT culture is based is:

> All employees must *always* look for the little things under their influence that they can do today to improve.

In the past, companies encouraged their employees to look for big improvements that could save large amounts of money; the little things were lost. But all those little improvements add up to big savings and big improvements.

A typical list of "little" improvements might include:

- Moving a bench to reduce the distance walked
- Ordering a tool to increase accuracy within a process
- Organizing tools for easy access, reducing setup time
- Developing a new rotation of jobs procedure to reduce boredom
- Cross-training people to do just one more job

Various types of suggestion systems have been implemented by companies. Unfortunately, while some are successful, others fail to achieve their goal of encouraging employee input.

Suggestion Systems That Fail

Hewlett-Packard had several employee-involvement programs during my tenure there. I remember those employees who won prizes for the suggestion that created the greatest savings. People didn't bother submitting smaller suggestions, believing they wouldn't win anything. Pretty soon managers were begging for input, and the programs all fizzled.

Hewlett-Packard needs very little introduction, so I'll just summarize the company as I experienced it. HP was known for high quality and leading-edge technology in computer products and high-tech electronics. During my tenure there, the company realized that more was required to remain competitive—it must also become a manufacturing company with emphasis on marketing.

I think the key to HP's success was a culture that allowed and encouraged innovation naturally—not through a formal win/lose suggestion system. It was okay to try new ways of doing things—it was even okay if the ideas were not perfect. Communication was open, and the "levels" were pretty transparent. HP was a fun and rewarding place.

Another company reported that it received 265 suggestions per employee during a particular year, of which 97 percent were implemented. I am sure that all of these "small" suggestions yielded far more savings than the few "big" ones that are captured through most suggestion systems that reward the big ones as winners.

Institutionalizing Only Value-Added Activities to Eliminate Waste

It is important that a company seeking employee input emphasize the need for ways to eliminate waste.

What is waste? It's anything that does not add value to the product. Counting, moving, extra transactions, unnecessary inspections, rework, storage, and waiting are all waste. As my colleague at Rath & Strong, Ed Hay, defines value-added activity in his book, *The Just-In-Time Breakthrough,** "Only an activity that physically changes the product adds value." If a certain activity is difficult to classify as value-added or waste, ask the question,

> "Will the customer pay more if this activity is performed?"

For instance, most parts are counted every time they change areas, and often a part is counted ten times in a process. Will the customer pay more because the part was counted ten times instead of one? Of course not; the part is priced on the assumption that the count shipped to the customer is accurate.

Who pays for all these wasteful activities? They add cost to the product, either raising the price or lowering the profit margin. In the end, usually the customer pays, which probably costs a company market share in a cost-competitive market.

A value-added analysis is performed by documenting every activity within a certain process—not by looking at process documentation but by asking people what they do. Each activity should be stated in a clear, concise manner. Then each activity can be rated as value-added or waste.

Most companies find that less than 17 percent of all activities add value to the product. Approximately 35 percent of nonvalue-added activi-

*New York: John Wiley & Sons, Inc., 1988.

ties can be eliminated quickly and easily, and there are many more opportunities for additional elimination in a more long-term effort.

Three Hints for Conducting a Value-Added Analysis

1. Have one person who does not work in the area involved in the analysis. This can be an outside consultant or a person for another process area or function. "Outsiders" will tend to ask the "dumb" questions to get all the process steps; people in the area take many steps for granted and will forget to document them. People who work in the area may also be hesitant to document those steps that are not "legal"—all those activities they added to make the process really work.
2. The analysis is best done by people working in pairs—again, to make sure all steps are documented. The combination of one outsider and one person who is familiar with the area is ideal.
3. If there is a question about whether a particular activity is value-added and the analysis team cannot resolve it, count it as value-added. There will be plenty of other nonvalue-added activities to eliminate.

Once the analysis is complete, the war on waste begins. First, rank order the nonvalue-added activities for elimination. Then those high-priority items to be eliminated should be assigned to those who are best able to eliminate them. Everyone in the organization can now begin to question all activities for possible elimination.

Exhibit 1-1 (at the end of the chapter) shows a value-added analysis from a manufacturing area in AGFA Compugraphic, a division of AGFA Company, a $900-million operation. The division produces graphics input and output devices in Haverhill, Massachusetts, and Wilmington, Massachusetts. Notice that only 21 percent of the sixty-six steps add value to the process of building keyboards.

Making Quality Everybody's Job

The goal is 100 percent quality. Customers demand it. Survival as a successful organization depends on it. As soon as quality gets close to 100 percent, the measurement of quality should be changed from a percentage to parts per million in order to keep the focus on improving quality. Quality is everybody's job. It is an integral part of any successful Just-In-Time implementation. The idea is to not only guarantee the customer 100 percent quality but to build only 100 percent quality (a detailed discussion of quality appears in Chapter 4).

Problems are exposed and attacked in a JIT environment. The focus is on solving problems once and for all by getting to the root cause. Short-term fixes cannot be tolerated. In the past, many problems have been hidden by extra inventory "just in case."

The rate at which problems are solved determines the rate at which more problems can be exposed by reducing inventories. JIT is *not* an inventory reduction program—instead, inventory reduction is a result of successfully solving problems (an important concept that is expanded on in Chapter 8).

Using Simplicity and Flexibility to Achieve Results

Just-In-Time's emphasis on making the right thing at the right time to satisfy the customer forces people to focus on two themes:

1. Simplicity
2. Flexibility

As an example, assume that an engineer shopping for a new piece of equipment to perform three operations walks into an equipment showroom and is met by a salesperson at the door. The salesperson says, "Let me show you an amazing piece of equipment. It does ten different processes with only the push of a button. Besides that, it runs at the speed of light. This machine costs $1 million, but it's worth it." The engineer is amazed. Looks like a good investment. After all, the capabilities are almost endless.

The salesperson then leads the engineer over to a small table-top model about the size of a microwave oven and says, "This smaller, simpler model does only the three different processes you require. It doesn't run at the speed of light, and you must push three buttons to run it. It costs $10,000."

Which is the better buy?

The engineer's requirement is to support five different production lines—none of which run at the speed of light. At the lower price, a machine could be purchased for each line—even a spare is possible. What if the large machine goes down? All five lines are stopped. What if one of the small machines goes down? To satisfy the requirements of simplicity and flexibility, the small machines are best—not the high-priced, high-tech one.

The primary requirement is supporting a process. In the past, success was dependent on how well a company developed reactive "fire hose" skills. Today process-improvement-for-the-long-haul thinking is required.

Learning From Those Experiencing JIT

One group leader at Compugraphic has said of JIT: "The concept is all-encompassing. Communication among all groups is a very important element of the concept, and all groups must understand the company's goals and how they relate to them."

Valleylab, located in Boulder, Colorado, is part of the Hospital Products Group of Pfizer, Inc., and produces electrosurgical products, both reusable and disposable, with sales of about $80 million. At Valleylab, one employee responded to the question, "What does JIT mean to you?" with the following: "Increased productivity, efficiency, cost control, and quality, along with an increased sense of pride in being part of a team. . . . Working together to save time and money for a better-quality product."

Building a Successful JIT Implementation

About three years ago, I began to formulate the "Building Blocks for Implementing Just-In-Time" concept as a way to help companies better understand JIT and develop an implementation approach for JIT concepts and techniques (see Figure 1-1). The building blocks analogy presents a sensible, organized way of explaining not only the elements of Just-In-Time but also the interdependencies and timing issues of implementation.

Imagine a child stacking blocks in a pyramid. What happens if the bottom center block is removed? Of course, the pyramid falls down. A Just-In-Time implementation is much the same as a pyramid of blocks. If the bottom middle block—People—is taken away, the pyramid, or implementation, will fall down. The People block is the cornerstone of JIT's foundation.

Culture and commitment are "people" issues. If these issues are constantly addressed in a correct and thorough manner, a company can build a JIT implementation with the cornerstone laid firmly in place.

Two additional blocks complete the foundation. Those blocks are Quality and Drumbeat. A Just-In-Time implementation is ineffective without Quality; JIT and Quality are like two sides of the same coin. Drumbeat includes planning and scheduling, plus basic business issues such as inventory accuracy, bill-of-material accuracy, and capacity planning. The goal is to create a regular, linear schedule (or rhythmic drumbeat) from supplier to manufacturer to customer.

People, Quality, and Drumbeat make up the foundation for every manufacturing company, regardless of its environment, products, or processes. Service organizations can also benefit greatly from these ideas; the product does not have to be one that can be handled. Administrative

Figure 1-1. The building blocks for implementing Just-In-Time.

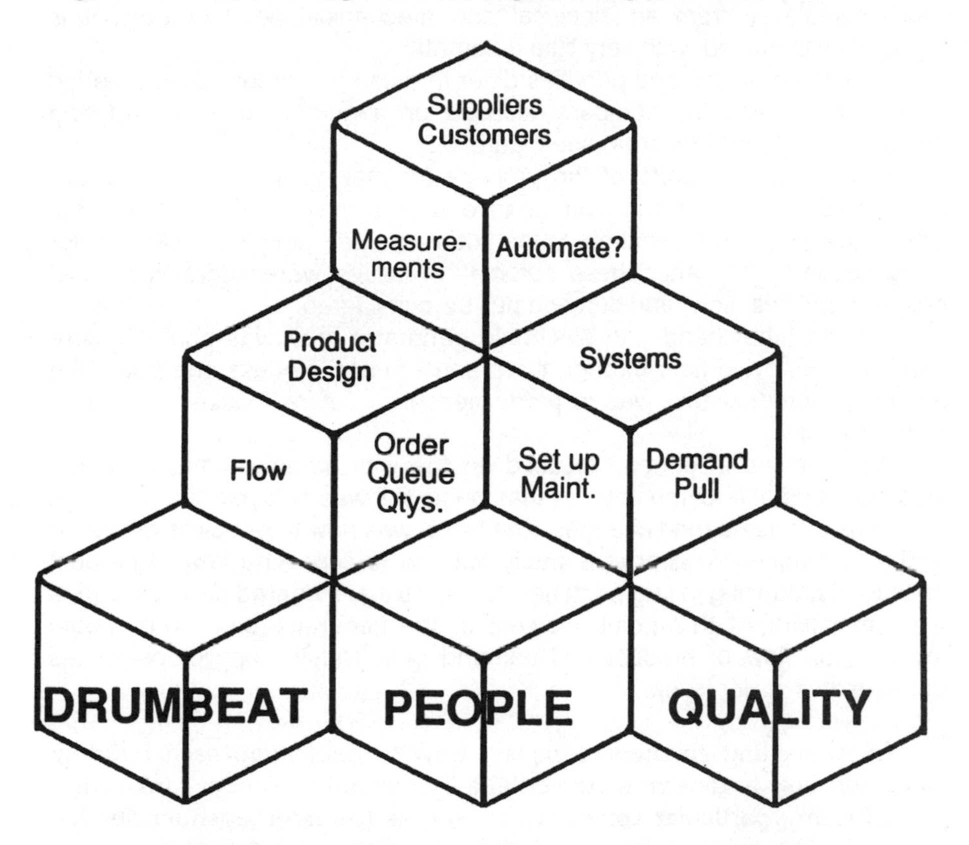

functions such as order processing or accounts payable can also implement these concepts and techniques.

The other blocks in this pyramid represent issues both inside and outside the factory. These blocks may be stacked in different priority sequences, depending on the particular environment, product, and process.

Setting Priority Sequences

At Valleylab, Inc., there are two very distinct process areas within a single division. One product line is surgical pencils (mostly disposable), built in very high volume with automated and manual assembly—a fairly simple product by some standards. The other product line builds the electronic generators that drive these pencils. This is a very low-volume product with a great deal more complexity from an electrical and mechanical point of view; it is manually assembled, with very little automation.

Surely the issues and priorities differ in these two areas. Over a period of several months the company focused on a number of manufacturing issues for each product area (see Figure 1-2).

Because some parts of the process for making surgical pencils are automated, equipment and maintenance were a major issue for that area. The supply of parts compatible to the automation at hand was also a major consideration. Only after these automation issues were addressed could queue quantities, flow, and demand pull be considered.

On the other hand, the electronic generator area was primarily concerned about cycle time through the manufacturing process and about the amount of product that was in production; it therefore focused on queue sizes and flow.

As Valleylab in general focused on the way manufacturing inventory was managed, it became obvious that much of it was delayed due to design problems or engineering changes. The issue was how to get parts delivered to the manufacturing floor in a timely manner to support a flow of product instead of producing in large batches. Systems that triggered parts deliveries were modified. At that point, workers in the generator area could begin focusing on flow of product and demand pull. Today, supplier parts has become their major issue.

In succeeding chapters, I explain how to implement each building block and how to determine the criteria by which to stack, or rank-order, them for any particular company, as well as the interdependencies be-

Figure 1-2. Building blocks arranged for two product areas.

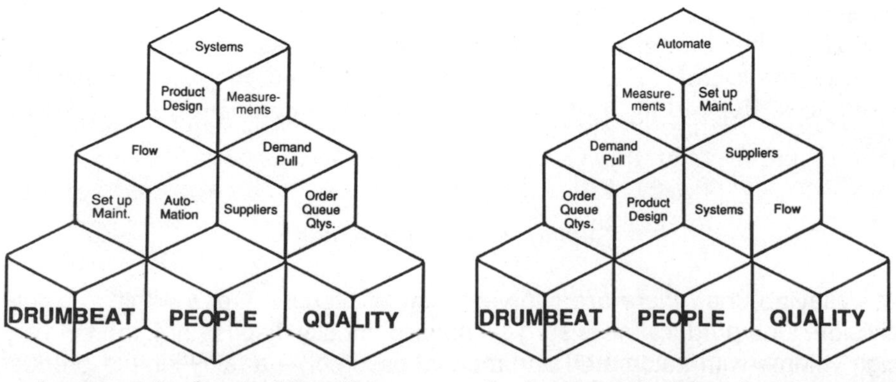

a. Arranged for Surgical Pencils b. Arranged for Electrical Generators

tween different blocks. First, though, there must be a clear vision of the company goals that is understood by everyone involved.

Creating a Clear Vision of the Future to Ensure Success

What would happen if I said to a group of people, "Let's all get into our vehicles and meet this afternoon at 3 o'clock"? I provide no map with a designated destination. Of course, the people would all leave, travel, and end up somewhere—probably their favorite fishing hole. However, if a map is distributed with a specific destination and with certain boundaries for travel, the people will head toward the destination. Now, some will drive their car, while others will ride in their truck, on their motorcycle, or on their bicycle. But since the destination has been clearly defined and understood, everyone will get there by 3 o'clock.

Similarly, if a company's vision, created by top management, is clearly defined and understood by members of the organization, all efforts are more likely to move the company closer to that vision. I refer here not only to the marketplace and technical work but also to the state of mind, or climate, that will prevail. To create a clear vision of what a business organization should be like in the future under JIT, two things must happen.

1. Management must develop a clear understanding of JIT concepts and techniques and must communicate it to all members of the organization.
2. The organization must then extend these JIT concepts and techniques into its own culture, management style, products, and processes in a way that will best meet customer requirements, creating a company vision for the future. This is a vision not just for manufacturing but for the entire organization, including the administrative, or indirect, areas.

It is perhaps more difficult to communicate that vision to every person you manage than to create the vision. In order to do this effectively, you have to know what the vision is, buy into that vision, feel that it is "doable," and, finally, know the role of every person you manage and how that role is integral to fulfilling the vision.

Improving Through a Common Vision

Management cannot just talk about this new vision; it must demonstrate clear commitment through action. Even though each company's vision is different, there are some elements that are common to all companies embarking on creating a Just-In-Time culture. They include:

- Employees who think and participate
- Employees who work smarter, not harder
- A team orientation
- Managers and supervisors who facilitate the work of those they supervise rather than dictating policies and procedures
- Process improvement as a major performance measurement
- Flexibility
- Focus on attaining a daily schedule every day
- Flow, balance of work
- Producing at the optimum speed, not as fast as possible
- Effective production—that is, doing the right thing first
- A quality emphasis that requires building quality in, not sorting out the bad ones

These elements suggest that people at all levels will be more involved in decision making and problem solving. They will be given more freedom to decide how best to move toward the vision within some reasonable boundaries. More than ever before, people must be given a greater feeling of power—a "sense of empowerment"—by management.

Once people have become involved and have experienced some successes, they can begin to develop ownership.

It is unrealistic to ask people in the organization for commitment before they have had a chance to be involved and develop ownership, and the only way to make reality out of the vision is to develop commitment from everyone to change the culture of the organization in certain ways.

INVOLVEMENT = OWNERSHIP = COMMITMENT

People closest to a problem situation generally have the best ideas and solutions. Time and again production workers have solved problems in just a few weeks that management and technical people have worked on for years. Empowering everyone at every organizational level requires every person in the organization to ask, "Can a person at the next level down, closer to the immediate issue, make this decision or solve this problem?"

Developing a Sense of Ownership

On a particular product line at Valleylab, the engineers and managers designed the process, arranged the line, and trained the manufacturing people on how they intended the product to be built. The workers came in each day and built product as quickly as possible per the specified process. Since they were not involved in the process design, they displayed very little ownership for making it work. Instead, there were always excuses why it wouldn't work—an "It's not my problem" attitude.

Later, the workers on this line were asked to develop a new process design for the line. This followed some very specific training. The workers designed a process that had many improvements; with engineering's help, they were able to implement the changes. The design still wasn't perfect, but now the group felt ownership for making it work—"It's not my problem" turned into "What can we do to improve this and make it work better?" As the workers improved their own design, a new commitment developed from the successes they were experiencing.

In reality, a company's vision is developed and fine-tuned over time through experience. Most companies change in small, incremental steps using pilot efforts.

* * *

JIT tends to be much more of an evolution then a revolution. Cultures don't change overnight. Gradual, consistent change in the right direction is best if you have the time. However, in a survival case such as Harley-Davidson, fast and dramatic changes must occur. In either case, to make these changes, the vision of where the company is headed should be clear and compelling.

Exhibit 1-1. AGFA Compugraphic value-added analysis.

PRODUCT: Keyboard

Step #	*Element*	*+/−*
1.0	Kit arrives at Drop Zone	−
2.0	Wait	−
3.0	Kit Breakdown	−
4.0	Move to #1 Assembly	−
	NOTE: Kit requires two carts—move one at a time.	−
5.0	Wait	−
6.0	Move keyboards to Disassembly	−
7.0	Wait	−
	NOTE: Balance of Kit parts wait #1 Assembly until the Disassemble Operation is completed for the entire Sales Order	
8.0	Disassemble	
	NOTE: The following operations are performed on each unit and all units are completed prior to move/return to #1 assembly.	
8.1	Move keyboard unit to bench	−
8.2	Unwrap	+
8.3	Remove card stock/discard	+
8.4	Flip	−
8.5	Remove 3 screws	+
8.6	Disconnect 3 connectors	+
8.7	Remove IC	+
8.8	Pry off board	+
8.9	Move board to bin	−
	NOTE: Bins are not of a standard size—they generally hold 30 boards of this type.	
8.9.1	Wait for full bin	−
8.9.2	Move full bin to #1 assembly area	−
8.9.3	Wait for remaining boards	−
8.10	Move keyboard to cart	−
8.10.1	Wait for remaining keyboards—S.O. balance	−
8.10.2	Move complete cart to #1 assembly area (−)	−
8.10.3	Wait for operator (−)	−
9.0	Desolder/Solder Boards	+
	NOTE: The following questions are performed for each board—all boards are completed prior to any further keyboard assembly.	
9.1	Move board to bench	−
9.2	Desolder	+
9.3	Remove old parts (2)	+

9.4	Position new parts (3)	+
9.5	Solder	+
9.6	Move board to cart	−
9.7	Wait	−
10.0	Reassemble keyboard	+
11.0	Move to line	−
12.0	Wait	−
13.0	Move to test	−
14.0	Wait	−
15.0	Move to test table	−
16.0	Test—100%	−
17.0	Record test	−
18.0	Return to line	−
19.0	Wait	−
20.0	Move to Inspect	−
21.0	Wait	−
22.0	Move to Inspect station	−
23.0	Inspect	−
24.0	Log Inspect	−
25.0	Return to line	−
26.0	Wait	−
27.0	Move to Base Assembly	−
28.0	Wait	−
29.0	Assembly (on-line)	+
30.0	Move to Audit	−
31.0	Audit (2 of 15)	−
32.0	Move to Pack	−
33.0	Wait	−
34.0	Clean and Inspect—Exterior	−
35.0	Wait	−
36.0	Add labels	+
37.0	Wait	−
38.0	Inspect—Visual	−
39.0	Wait	−
40.0	Log Inspect	−
41.0	Pack in box	+
42.0	Write Part Number on box	−
43.0	Move to skid	−
44.0	Wait	−
45.0	Move to Pick-Up Area	−

	#	%
• Total Elements	66	
• Add Value	14	21
• Add Cost Only	52	79

Chapter 2

Expanding Your Resources to Succeed With JIT

The People block is the cornerstone in the pyramid of building blocks for implementing Just-In-Time. If this building block is laid firmly into place, any company can be successful with JIT. *People* refers to all employees in a company, from the chief executive officer (CEO) to the hourly workers. Within an organization, people are classified into "managers" and "workers." Anyone who manages people on any level is a manager, whether a supervisor, a group leader, or a senior vice president. A worker is anyone who has a manager. Many people in large organizations play two roles—that of manager and that of worker.

Worker involvement and empowerment of workers at all levels push decision making and problem solving down in the organization to the people closest to the immediate root-cause problems.

Most companies today look like Figure 2-1a. There is a very broad scope of involvement at the top-management level, while the scope of activity of most workers is very narrow. Many people in such organizations feel they are at the bottom of the triangle and that they have very little influence or empowerment in their company. As the training manager at Compugraphic recently told a group, "It's like the shopping mall directory—You are here."

The challenge is to reverse this triangle, as in Figure 2-1b, so that workers have a broad scope of work and the influence to make things happen. Management is then able to focus on strategic planning and global issues.

As this change occurs, management at all levels no longer has what one Valleylab supervisor called "the burden of having to know everything and of being the solver of all problems"; this now becomes a shared burden and therefore, a lighter one. As companies make this transition,

Figure 2-1. Scope of involvement and influence in traditional hierarchy (left) and in JIT (right).

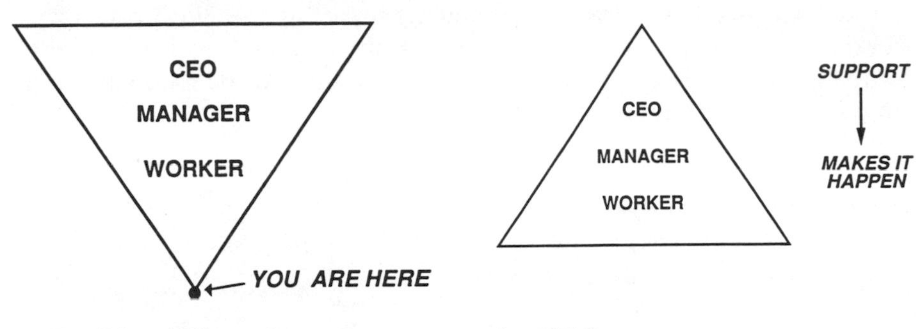

a. Traditional Hierarchy b. JIT System

they discover in their people a gold mine of knowledge, ideas, and solutions.

You, as a leader, will find the challenge of balancing "letting go" and "controlling" a difficult one at best. Strength, confidence, and sensitivity are required from leaders to encourage commitment to continuous improvement. An effective leader of this transition contributes to a culture that supports change and creates a "can do" attitude throughout the organization. How well companies tap this resource is a key indicator of success as they implement JIT concepts and techniques.

In this chapter, I discuss the benefits and challenges for managers and five attitude changes for managers, followed by other organizational issues including maintaining job security, handling unions, timing the change process between managers and workers, and dealing with those who don't want to get involved.

The Benefits and Challenges for Managers

Involving and empowering workers can be very positive for managers. However, it can also be threatening—especially to mid-level managers and supervisors.

Using the skills of individual workers to their fullest and getting workers to deepen and broaden their skills not only help the organization as a whole but also help each individual manager do his or her job better. There are three benefits for managers of an involved, empowered work force:

1. Time to plan ahead for the future
2. Ability to proact instead of just react (fire fighting)
3. Opportunity for career development, e.g., seminars, classes, books

In spite of these and other benefits, getting middle management buy-in is one of the toughest issues most companies face. Middle managers can feel squeezed between top management forcing decisions down and workers now forcing decisions up. Recently, a group of leaders from companies experienced in implementing JIT identified the following challenges as critical issues for middle management:

1. Worries over job security
2. Lack of training
   - Awareness in general
   - Deeper education of process
   - Implementation
   - Human relations
3. Attitude of "I've always done it that way"
4. Failure of top management explicitly to ask middle managers to make and fulfill commitments
5. Inconsistent top management message/measurement systems that still reward supervisors and middle managers for meeting quantity targets
6. A sense that the "rules of the game" have been changed midstream
   - Became managers by virtue of technical knowledge
   - Are now asked to be "coaches" and "facilitators"
   - Have no sense of real power or influence

In order to counteract these possible blocks, companies need to clarify the role of supervisors and middle managers in a JIT environment right from the start. This can be done by clarifying the potential job restructuring that will be undertaken and training supervisors/middle managers in teamwork, problem solving, interpersonal skills, performance management, and goal and strategy formulation. Too often the performance of middle managers is measured in indirect ways only periodically; day-to-day performance measurements must be created that encourage doing things the right way.

Top management must have a consistent focus that is visible, vocal, and viable. Middle managers must be brought on early, involved, and shown how they fit into the overall picture of success.

How to Motivate Your Employees to Achieve Success With JIT

The five changes in attitude for JIT success include:

1. Shift from being an authoritarian to being a facilitator; work through people instead of just working people.

2. Make different assumptions about workers, their cares, desires, and abilities.
3. Decide that it is all right to make mistakes, as long as those mistakes become learning experiences and enhance long-term improvement.
4. Realize that in a JIT environment the worst evil is not stopping the line; rather, the worst evil is making bad product.
5. Replace the "quantity, quantity, quantity" mindset with one of "quality, quality, quality."

Breaking Those Old Perception Barriers

In order to successfully change attitudes, paradigms that have been created by generations of American manufacturing tradition must change. A paradigm is a mental model that represents how things are or how they ought to be. In the United States, manufacturing culture has created a paradigm about how manufacturing ought to be done, and it is very hard to change how people see their role in the manufacturing world and how they see the dangers and potential of change.

When a person seems slow to change, it is not always because of stubbornness; the person may have internalized a paradigm that governs how he or she sees the world. The person simply cannot see how change will be better, at least not until he or she gets more involved in the process and has an experience that empirically breaks the mental mold that has been developed.

It seems that a different experience may be required to break a paradigm. Many times, paradigms are broken by "new" people or outsiders.

Breaking the Paradigm From Outside

At AGFA Compugraphic, a worker team was working on a goal to improve the flow of a product through a particular process. One piece of equipment required was in another area and could not be moved because that area also used the equipment. The team determined that if it had this equipment in its area, the flow would be much better and time wasted walking back and forth would be eliminated. The team identified $200 per year savings in time. Frustration and aggravation, not easily quantified, were not factored in. A new piece of equipment cost $3,000. The team hit a dead end; a $200 savings did not justify a $3,000 purchase. The team felt defeated, but still the members wanted to do something.

After a few weeks, a person outside the team broke this deadlock by suggesting either purchasing used equipment or a smaller unit or building the equipment from excess material. Someone else thought about communicating the group's need to the entire facility to see if anyone else had a solution. The result was that they were able to locate another group of people who had the equipment but were not using it.

The paradigm here was that if a piece of equipment was needed, it had to be purchased, and it had to be new. By breaking that model and seeing that equipment can be built, bought used, or found lying idle in another part of a facility, the group was able to look at its entire world of work with different eyes.

In order to get the most out of each and every worker in a facility or a company, look through different eyes at the entire world of work in which you and the people you manage function. Breaking paradigms about how relationships ought to be and how people ought to behave and what they ought to expect from a worker will give you a new lease on life to begin improvements. Now you are fired up to do what makes sense—not what you have always done.

How Your Becoming a Facilitator Can Help Motivate Your Employees

Becoming a facilitator can help tap that "people" resource in your organization.

The authoritarian manager says:

"You work, I think."
"Build as fast as you can."
"The inspector is responsible for quality."
"Build only what I tell you when I tell you."
"Specialists and superstars get ahead."
"Always stay busy."

An authoritarian advocates and enforces unquestioning obedience to, or compliance with, authority. It's a "do as I say" environment.

I asked a group recently what the word *facilitator* meant to them. People replied *coach, supporter, helper, teacher,* and *idea catalyst*. They understood! A facilitator works through people while an authoritarian works people. A JIT culture requires that everyone participate and think. Becoming facilitators may appear to be easy for management—most managers would love to have less detail work and more time to plan and think strategically. However, this transition can in fact be quite challenging and unsettling, especially if *you* are not prepared for the new role.

You may feel insecure as a facilitator, wondering what your new role really is. Workers are taking over much of the decision making and problem solving that always rested with management; in fact, the ability to solve problems well is what won most people their management positions. The transition from an authoritarian style to a facilitative one is like going from being a commander to being a coach. Coaches still have the final decision-making authority, but their job is to instill into the team members the ability to think and implement solutions themselves.

The role of a facilitator requires a lot of "people skills" that have not been necessary in the past. The role does not require that you no longer make decisions or assume control when necessary; the difference is in how you exercise your power and when you do so. Managers who have made a successful transition rarely have a need to exercise their power, because well-educated people who feel ownership and empowerment will usually make the right decisions given enough information.

A facilitative manager says:

"Try your new idea."
"Build what is needed to meet the customer need."
"Everyone is responsible for quality."
"You know best how to get the job done. Let me know how I can help."
"The more different jobs that you can do, the better."
"If you are idle, help someone or work on improvements."

A part of becoming a facilitator is conveying "the big picture" to workers and making each individual's role clear in that big picture. In order for people to make sound decisions, they need to understand what part they play in the overall scheme of things within a company.

Changing Managers' Assumptions to Allow Workers to Solve Problems

As you develop the skills of a facilitator, you will solicit ideas from workers. But that is not enough. You then need to support and encourage the implementation of these ideas by those workers. In the past, workers often identified problems, then pitched them into management's lap for a solution. Management got a great deal of satisfaction out of putting on the fire hat and putting out the fire.

As a facilitator, you cannot accept a problem without some carefully thought out solutions from the workers. In order to do this well, you will assume workers have:

- Integrity
- Intelligence
- Common sense
- Trustworthiness

These management assumptions give workers:

- Security
- Confidence
- Respect

In every company that I have been associated with, the overwhelming result of management letting workers define and solve problems is that workers come through with flying colors. They truly care about their company. They want to do a good job, and most of them love having the power to create change and improvement. Photos 2-1 and 2-2 show how one team reorganized its area to store parts at work stations, eliminating wasteful movement and the chance of using the wrong part.

How to Encourage Innovation for Improvement

For new ideas to work, it must be acceptable for them to fail. If this seems paradoxical, consider this: Has anyone ever had a new idea that worked perfectly the first time? Probably not. If perfectionistic cultures are created in companies, new ideas will be scarce. When perfection is required in an organization, people become "thinkers" instead of "doers"—they are always planning more or analyzing "just one more thing" to be

2-1. This photo demonstrates the efficiency and good planning that teams typically employ when empowered to lay out their own work space. The worker has all the tools and materials she needs placed in open bins within easy reach, eliminating extraneous steps that waste time.

2-2. This photo shows the tools and materials the way they were before, on shelves away from the work area, an inconvenient location that wasted movement and time.

sure the change will work when implemented. People become extremely conservative and risk-averse because they are afraid of failure. In this environment, if someone does try a new idea and it does not create the expected results, the reply from the authoritarian manager is, "Now, see, I didn't think that would work. Change back to the way we've always done things and ship product."

Henry Ford said, "Thinking always ahead, thinking always of trying to do more, brings a state of mind in which nothing is impossible. The moment one gets into the 'expert' state of mind a great number of things become impossible."* A mindset of continuous improvement creates an environment of continuous change. People learn by trying new ideas, with the understanding that some results just cannot be anticipated no matter how well the situation is analyzed. People are willing to try new ideas when the fear that these ideas must work perfectly the first time is eliminated; managers who have made the transition to facilitator can promote and foster the idea that it's okay to make mistakes. The ultimate goal is to make things better. Many things actually seem worse than ever before an improvement is realized. In this new environment, if people try a new idea that does not create expected results, the answer from the manager is, "What did you learn from this experience? Good, now take this new knowledge and continue until you are successful."

*Henry Ford and Samuel Crowther, *My Life and Times* (Salem, N.H.: Ayer Co. Pubs., 1922).

You may feel very discouraged during this change process if it is not clear to everyone that most of the time things get worse before they get better during any improvement effort.

Notice in Figure 2-2 that this process dropped below status quo before crossing the line to benefit. The only way to have gotten the benefit was to drop below the line first. This represents the learning curve and experimentation that are needed for innovation.

In the words of one worker in a JIT company, "We should all try to be open to new ideas because without the willingness of people to experiment there would never be any progress."

Why Stopping Cannot Be the Worst of All Evils

What is the worst thing that can happen in manufacturing? What brings the plant manager down to the production department in a cold sweat when nothing else will? Of course, it's *stopping*. Managers in manufacturing seem to have a beeper implant that immediately goes off when production stops.

Stopping seems like waste—waste of people, equipment, parts, and all that overhead. Huge piles of inventory are traditionally built just to keep everything running.

Figure 2-2. Impact of change.

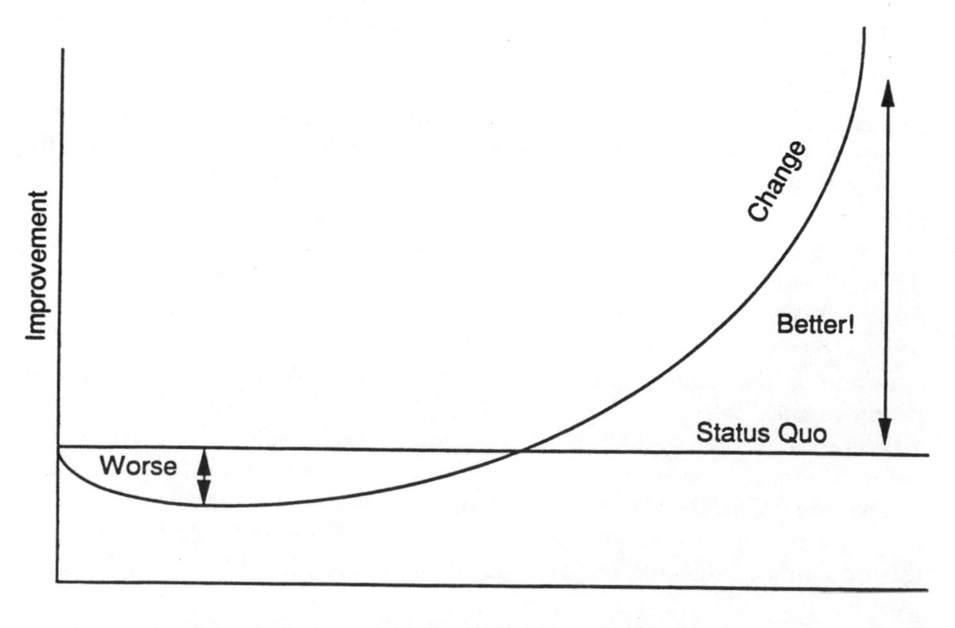

But building up inventory could create far more waste than stopping. Let's assume that there are three work stations in a particular production process—A, B, and C work stations. Work station C goes down (because of a parts shortage, equipment failure, absent operator, or any one of a number of other causes). Work stations A and B continue to produce, building up a huge pile of inventory just waiting for work station C. The plan is to work C overtime when it is up and running. Finally, C is running. A quality problem is found in the inventory. Now, instead of a few pieces to rework or scrap, there is a huge pile. This will cause further delays, rework and/or scrap. What a waste!

Changes in the design of the product or process are also limited when large piles of inventory exist. Again, lots of waste. Flexibility is limited. Parts may be used that are needed to complete other units. The cost, or risk, is not worth any perceived benefit. I'm not necessarily suggesting that if a line goes down everyone should be idle or go home, but workers should not build more just to keep busy. They may be able to help in another area, get some training, or take the time to have a team meeting without a time limit—a leisurely brainstorming session.

In a JIT environment, the worst of all evils is not stopping. The worst evil is making bad product. Next is having problems and inefficiencies that are hidden. The third worst problem is having the flow of production stopped.

Figure 2-3 shows that the worst evil in the traditional environment has been relegated to number three in the JIT setting. The biggest issue is not how to physically manage the work but how to change the mindset to say, "It's okay to stop for problems. In fact, it's clearly the only right thing to do." It doesn't seem logical that stopping for problems and solving them will improve the ability to meet schedule. But it works.

The first few steps are the most difficult. The workers cannot assume responsibility to stop without your approval. This is a very critical change in management attitude. I relate it to skydiving, since I am deathly afraid of heights. It would require a strong push to get me out of that plane. I would be terrified. However, eventually I would remember something about pulling the ripcord to open the parachute for survival. I would do so

Figure 2-3. Worst evils under different systems.

Traditional	JIT
1. Stopping	1. Bad product
2. Covering problems	2. Covering problems
3. Bad product	3. Stopping

and survive quite well. The second jump would be a little easier, and for the third, I might even be confident. Every time I jumped, it would be easier, because my experience, or involvement, would have created some ownership and confidence. Similarly, the first steps will be the most difficult for you when it comes to allowing work to stop.

One production manager I know understood that stopping was the right thing to do. However, the first time that production stopped for a problem, he was so nervous about making shipments that he simply left the plant. He knew that if he stayed, he would create so much tension that the production people might not take sufficient time to solve the problem. It took three experiences of stopping before he was able to stay. An amazing thing happened during that time—the plant still made shipments as planned. In addition, the problems were solved, not worked around. Stopping actually created less waste in the long run than building up inventory.

How Focusing on Quality Will Result in the Right Quantity

Management has always been forced to focus on quantity: "How many did you build? How fast did you go? Did you keep that expensive machine going?" This management attitude is a reflection of measurements based on "more is better"—even if more is not needed. Changes in measurements will be discussed more in Chapter 14, but for now let's just say that emphasis on building the right amount at the right time in a balanced and synchronized fashion will satisfy the customer's requirements better than a focus on quantity.

Focus on quantity creates a frantic, fire-fighting style of management. Most of us have been on a playground merry-go-round. As the merry-go-round starts spinning faster and faster, we can't think how to stop it. We only know that we're feeling very dizzy and sick. But once we finally get off the merry-go-round, we can easily devise a procedure that would have stopped it. A key to success is to "stop spinning" in order to solve problems, make things better, and actively manage the business. The only way to do this is to take away the emphasis on quantity and replace it with an emphasis on quality and building the right amount.

How to Deal With Four Key People Issues

There are a number of organizational issues that face companies looking to implement JIT. Two of the most important are the impact of JIT on job security and the implementation of JIT in a unionized organization. Others have to do with the timing of management and worker change and what to do when your people are not interested in greater involvement.

Maintaining Job Security

In a Just-In-Time environment, improvements in the process are constantly being implemented and the "war on waste" is in full force, eliminating nonvalue-added activities. People are working smarter, not harder. Productivity improves.

In a company that is growing, this is no problem. As more manpower is created through improvements, more product is needed as well. This simply means that not as many new people will be hired, but no jobs are lost. Those people with well-developed teamwork and possibly even facilitator skills increase their credentials manyfold. In every case I have seen, people in this situation have greater opportunity and job enhancements as a result of this corporate growth.

Other companies utilize JIT as a marketing tool. By increasing the ability to meet the customer's needs, companies can sell more. If a company can keep a stable work force until improvements increase sales, then any people freed up as a result of productivity improvements will be needed for more production. This balancing takes careful planning by top management.

But there is a third scenario that is less hopeful—that of a company that did not begin JIT "quite in time" to create more market through improvement. This company is experiencing low sales because of problems with delivery, design, or quality. Now, if some very firm decisions are not made by top management at the outset of a JIT effort, people may feel that if they make improvements, their jobs will be lost. This can be a show-stopper!

Even in a company in this situation, progress with JIT is not hopeless. However, special planning and some tough decisions must be made and communicated to all of the people before a JIT effort is commenced. *If people feel that by offering suggestions and making improvements they will eliminate their jobs, JIT cannot be successful.*

You can address these manpower issues early in a JIT implementation by following these steps:

1. Understand your marketplace and what it is going to take to grow in that marketplace. A carefully thought-out plan of goals with appropriate time windows for each will be necessary.
2. Determine the manpower (as well as other capacity issues) required for each stage of the plan. The outcome should be a guaranteed level of employment for a certain level of business. Adjustments and reorganizations should be done to bring reality in line with the plan.
3. Once a baseline employment level is established, people cannot lose their jobs because of improvements in the process. If business

conditions change over time and there is a downturn in sales that can be clearly stated to the people, then a reduction in work force may be necessary.

The key is to relate a reduction in work force *only* to an unexpected downturn in business—never to JIT or improvements by the employees.

If a reduction in work force (RIF) is necessary due to an unexpected downturn in business, the JIT mindset changes the traditional seniority system most companies use. Traditionally, people who have been employed by the company the longest are the most secure. JIT forces a company to look at contribution as the key criterion for security, not seniority. Seniority may be used as a tiebreaker, assuming equal contribution, but nothing more. Contributions such as process improvement, teamwork, and flexibility are strongly considered.

Getting Union Support for JIT

In a unionized work force, seniority rules are often strict, and seniority is one of the areas unions are loath to bargain away when negotiating with management about work-rule modifications necessary to implement JIT. Nevertheless, unions are not showstoppers for a JIT implementation.

However, a company with a union is forced to educate and plan more up front. The watchword is No Surprises. Union officials should be involved in every step of JIT planning. Most unions today agree that it is better to change than to shut down; it is better to have a secure future by doing things differently than to have no future at all. The key is to link having a successful future only with doing things differently, which requires that people be more involved and that they become more flexible.

No longer are people secure because they know how to run a machine better than anyone else. In the past, flexibility (running many machines) was not important in the shallow structure of each job classification. However, the number of classifications was high, limiting flexibility. Now it is the person who knows how to run multiple machines who is the most secure. This thinking requires that the number of job classifications be drastically reduced and that depth of knowledge, or flexibility, be added to each classification (see Figure 2-4).

One of the first steps in a JIT implementation involving unions is to train the key union people in JIT concepts and techniques. You should help them understand *why* JIT is key to future security and success and give them an opportunity to be involved in the planning as well as the execution of JIT.

Helping Management and Workers Change Together

Workers traditionally have looked to management not only for authority and decision making but also for expertise. Your workers expect you to

Figure 2-4. The relationship between job classifications and depth.

TRADITIONAL LEVELS	JOB CLASSIFICATION												
	1	2	3	4	5	6	7	8	9	10	11	12	13
Poor													
Acceptable													
Outstanding													

JIT LEVELS	JOB CLASSIFICATIONS		
	A	B	C
1			
2			
3			
4			
5			
6			
7			
8			
9			
10			
11			
12			
13			

already know everything about JIT as they begin. But the reality is that, more often than not, you are learning right along with the workers, experiencing a significant behavioral change along with them.

Persistence Pays Off

Recently, a production team on a final-assembly line experienced a parts shortage. Since the people on the line knew that balance and flow were good in JIT and building bunches was bad, they decided that they would not build up a bunch, wait for the part to come in, and complete the assembly. Instead, they would build the product in a start-to-finish flow when the part came in. They even thought of some ways to reallocate people so they could meet schedule.

When the plant manager heard about the new plan, he became very nervous. He told the team to go ahead and build the bunch without the missing part "just to be sure." The team proceeded to tell him why this was not the right decision; furthermore, it wasn't JIT.

The plant manager had the final say, and the bunch was built. The team was amazed that their manager was not following JIT behavior. However, they didn't give up, and the next time a parts shortage occurred, the team was allowed to wait until the part came in to build the product. They made the schedule by doing some careful planning.

The team learned two important lessons: One, don't give up—be persistent in the quest for change—and two, managers learn along with everyone else.

Dealing With the "You Don't Pay Me to Think" Attitude

Most people are very happy to have the opportunity to think and contribute. However, some would really rather just come in, do their job, and go home. Their response to the idea of being seen as a resource is typically, "You don't pay me to think." It is difficult for these people to see the benefit for them.

If these "not-so-excited" few are neutral in this process, continue to include them. One day some of them may be your biggest supporters; they need time to see to believe. Others will remain disinterested.

However, if there is a pulling against the effort, or negative energy, it should be made clear that involvement and teamwork are the only ways to be successful in the future. Negative activities cannot be tolerated.

Tapping this often untapped resource—people—in an effective manner is not easy. In fact, this is probably one of the least-understood pieces of Just-In-Time. However, total and long-lasting success with JIT requires this cornerstone block to be laid firmly in place.

Chapter 3

How to Create Teams That Effectively Unlock That Untapped Resource: Your People

One of the keys to worker involvement—a big part of the cultural change necessary to bring a company into a Just-In-Time environment—is the use of teams. The way in which teams are formed, educated, and facilitated is one of the most critical issues in implementing a successful, results-oriented, and permanent Just-In-Time environment. Individuals who have typically functioned independently require a new forum—teams—to learn how to work together. Involvement through team membership leads to ownership and, finally, commitment.

Most companies do not recognize the important role teams play in implementing Just-In-Time concepts and techniques. Efforts involving these interpersonal skills issues are usually overrun by emphasis on new techniques and quick bottom-line results. Taking shortcuts can yield some short-term success, but for long-term, continuous, significant success with JIT, working in teams is a must. It is only through the commitment of the people that implementation of new techniques will be successful over the long term. Most JIT techniques make sense—they do not require "rocket science" degrees to understand—but they do require an open mind and a willingness to experiment, something few companies have stressed to their work forces.

For example, when I began working with AGFA Compugraphic on a JIT pilot effort at Haverhill, the facility management didn't feel that the teams needed team-building training—after all, it already had "quality teams." As a result, at first less than full attention was paid to training teams. Within a year, however, more and more effort was being put into team building. One project team noticed that the two most recently formed teams had become productive, well-functioning units very quickly. A quick study was done to determine what training had been given each team in the plant, and the results were not unexpected: The two teams that were doing so well were the first to get *all* the team-building training.

In this chapter, I explain why teams are the mechanism for worker involvement. Then I discuss who should be on teams and the different ways of forming teams; development stages for teams; education and training requirements for team members and facilitators; and tips on managing team meetings. Last is a section on how to keep the momentum.

> One big caution for you as an implementor: Pay attention to interpersonal issues early—the effort will pay off.

Teams—The Sine Qua Non of JIT

Here are six reasons to use teams for JIT success:

1. *Encouraging participation.* Teamwork creates an environment conducive to employee participation. People feel secure as part of a group. Some of the most withdrawn people will participate if they feel part of a team.
2. *Generating more ideas.* People working together will always have more ideas than a group of individuals working alone. Team members encourage ideas from each other.
3. *Generating better ideas.* Ideas create more ideas. As one team member has an idea, that thought may spark a thought in another individual and so on, until a better and more thoroughly thought-out idea results.
4. *Developing a willingness to take risks.* There is a certain amount of risk involved in participation and idea generation—risk of failure, risk of looking silly, risk of sharing thoughts with others. There is less risk for a team than for an individual trying to make change occur.
5. *Fostering feelings of power and influence.* Feelings of power and influence can be obtained only through experience. People must believe they can make a difference. There is power in numbers, so a team more easily feels that it can and will have an influence on the organization.

6. *Improving quality of work life.* Innovation flourishes in a team environment, and once a team has matured, a high level of satisfaction is achieved. People experience more pride as part of a group.

Making Managers More Effective Through Teamwork

Teamwork is important at every level of the organization. A management team can be much more effective than a group of powerful individuals making individual decisions and negotiating with each other one on one. When managers have an opportunity to function in a team, consensus can be achieved and the best decisions for the company can be made.

Have you ever heard a manager say, "I don't agree with this decision, but we must follow the procedure anyway"? How committed are the employees in carrying out this manager's orders? As little as possible, of course. When people don't buy into a process, they are seldom effective in implementing it; the first chance they have, they will find a way around it. If this manager had been involved in the decision-making process—in a team—commitment to success could have been achieved.

"Staff" is a hierarchical term referring to those who report to a manager. A "staff meeting" is run by the manager in a report-and-record fashion—no time for consensus or opinions there. The top manager typically sits at the head of the table and tells the staff things, receiving updates from members when necessary. This "boss and direct reports" group can be referred to as a "natural work group."

However, as management teams consisting of a group of managers or a manager and people reporting to that manager are formed, the traditional "staff" and "staff meeting" evolve into more egalitarian forms. In trying to implement something as different as JIT, it is necessary for people to have a common vision of what things will be like in the new environment. And it's very hard to create a vision through this command-and-control structure. A team is far more effective in creating that common vision. A management team meeting is facilitated by a member of the team, not necessarily by the manager. Some management teams have been successful in rotating the facilitator role over time. Openness and honesty are encouraged. Understanding of each other's concerns is gained; the highest level buy-in comes from a team decision.

Meetings should become more like working sessions and less like on-stage events. Managers shouldn't need to have meetings to prepare for *the* meeting any longer.

Setting the Stage for Teamwork

Teams are important for success at every level of the organization. It will be difficult to form effective teams at lower levels if management does not set a good example at the top. If teams are working well at lower levels, generating ideas and solving problems, and top management is still functioning in an authoritarian manner, middle management will be getting pressure from above to "do as I say" and pressure from below to "let go and allow us to solve these problems." There will be a crash site somewhere in the middle (Figure 3-1).

In a JIT environment (Figure 3-2), communication flows freely in all directions. Listening becomes a very important skill at all levels.

In a company organized according to JIT, the structure of teams depends on the task at hand, not the level within the organization of the team members.

Two Types of Team Structure: Process and Task

Teams in today's successful JIT environment differ from the traditional quality circles or quality teams. Quality circles were typically composed of individuals who worked together day in and day out. Their charter was to uncover, or highlight, problems; it was then up to management to solve these problems. Since follow-through was poor and the teams had no power to change things, quality circles were generally not successful, at least not when used in American companies. In Japanese companies,

Figure 3-1. Pressure on middle management in traditional environment.

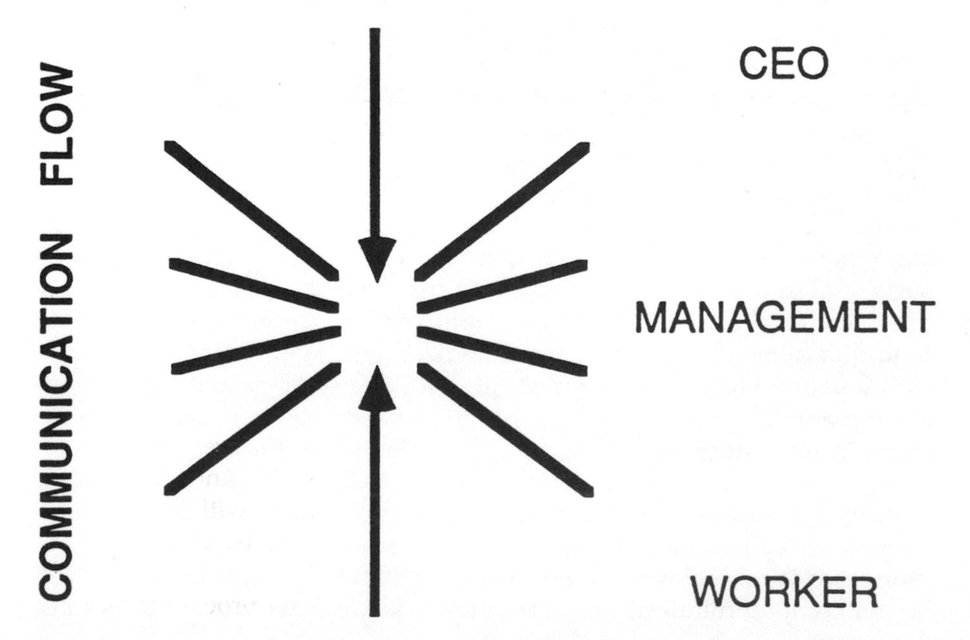

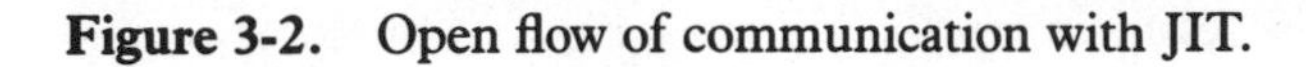

Figure 3-2. Open flow of communication with JIT.

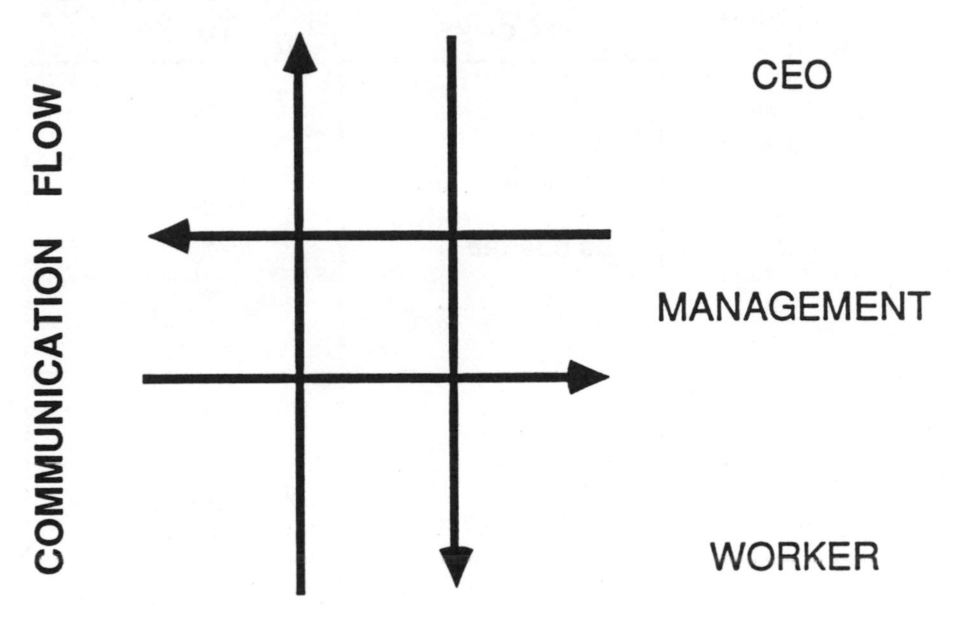

where industrial engineering is often the power behind the management throne, quality circles were often able to succeed because when a problem was discovered, the industrial engineers went right to work to solve it. But in companies in other countries, teams must be "empowered" to solve problems if they are to be fully effective.

As shown in Figure 3-3, there are two structures of teams in a successful JIT environment: process-oriented teams and task-oriented teams. These teams differ by:

1. Who the members are
2. Whether membership is voluntary or mandatory
3. How projects are selected for the team to work on
4. How long the team operates

How Process Teams Function to Improve the Process

Process teams include everyone in a particular work area, plus some support people as needed. Membership is mandatory, and the team is ongoing. The team's charter is to improve the process within its area of work. The team usually has freedom to choose projects, as long as the projects support the overall JIT goals.

In order to fulfull its goal of improving a general process, a process

Figure 3-3. Two types of team structure.

	Process	Tasks
Who	Work area and support groups as needed	Various functions
Membership	Mandatory	Appointed voluntary
Project selection	By the team within clear vision and goals of organization	By management problem–improvement technique
Team continuation	Ongoing	Disbands when task complete

team needs to understand clearly the vision and goals of the organization, as well as the vision of how the particular process will function in a JIT environment. The group will require a lot of guidance and clear goals from management, particularly in the early stages, to determine how to proceed.

One part of the vision for many companies is to reduce manufacturing lead time. With a goal of reducing manufacturing lead time by 50 percent in three months, for instance, the team might take any one of several roads. It may decide to work on creating work cells, better line layout, or better tooling or on reducing queues to begin to accomplish the goal. The key is that the team is involved in deciding which road to take to get there, as illustrated in Figure 3-4.

With freedom to pick issues to work on comes both challenges and advantages. The challenge is that a team may lose its way or perceive that it has run out of things to work on after functioning for many months or even years. The advantage for process-oriented teams is an underlying ownership for the issues they have chosen to work on; these are issues that have most likely been of great concern to team members for a long time; now, as a team, these workers are finally getting a chance to solve them. The team should stay focused on the goals to be achieved that will satisfy the business's needs.

A Process-Oriented Team at Work

One team at Valleylab works in the areas of receiving, inspection, and warehousing, covering the process of getting material from the dock to the production line. One of the goals given to this team was to eliminate

Figure 3-4. Process team approach to a problem.

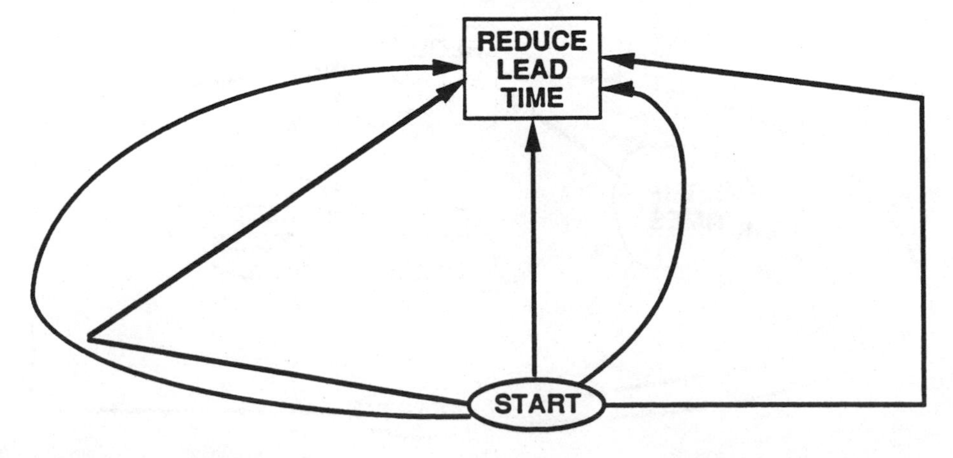

nonvalue-added activities in the indirect functions. The team chose to work first on the issue of time cards because, while it may seem a minuscule point, time cards have bothered many members who are required to process a time card every day despite the fact that time is not charged to specific jobs. A representative of the team was sent to accounting and was told, "You can't use weekly time cards; corporate won't allow it." But the team didn't want to give up; it asked a staffer from human resources to attend its next meeting and explained the savings that would be associated with a weekly time card. Soon the request was approved by accounting, and the problem was solved. Clearly, the team's determination to solve this issue was related to members' ownership of it.

How Task Teams Function to Solve Problems or Learn New Skills

Task teams are usually made up of a group of people, either appointed or volunteer (within guidelines), from a number of functions within the organization. The team is constituted to solve a particular problem, work on a particular improvement, or learn a particular new technique. When the task is complete, the team disbands.

Let's go back to the goal of reducing lead time and see how a task-oriented team might become involved. Instead of allowing the team to pick the route to the end, management defines a specific task (in this case, reducing setup time), and the team must follow a predetermined course.

The particular task illustrated in Figure 3-5 is setup reduction. A typical setup reduction team may be made up of an engineer, a planner, two setup people, and one operator. When the goal of reducing setup time by 75 percent is reached, the team will disband or at least be reformed to

Figure 3-5. Task team approach to a problem.

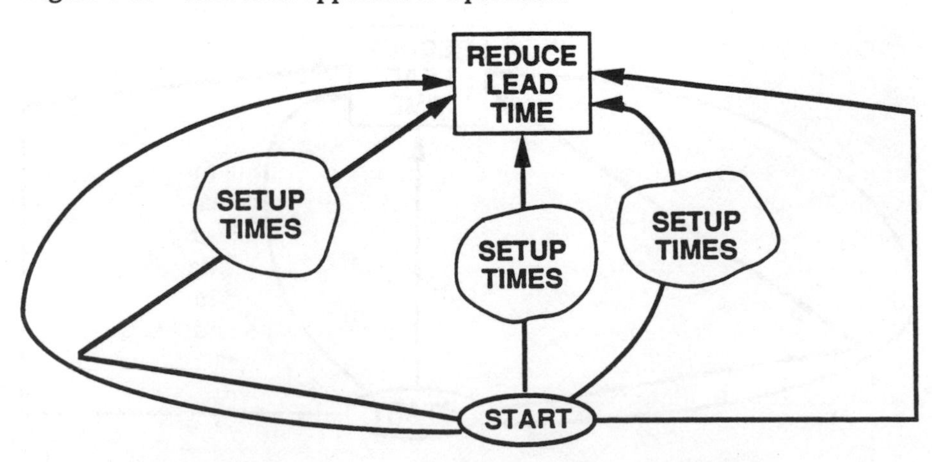

reduce setup time another 75 percent. In this example, a task team was used to learn a new technique—setup reduction. In other instances, a task team could be formed to solve a particularly challenging problem or to work on some other specific improvement.

Both Team Types Are Needed for JIT Success

I recommend having both types of teams in any JIT implementation effort. The key is to do what makes sense. Sometimes a process team is needed; at other times a task team is required to get the job done. Ensure that the way teams are structured and chartered fits with the business needs of the company. I strongly encourage a goal of having everyone in the organization involved in at least one team; this will help accomplish the cultural change necessary with JIT.

One of the most difficult roles you may be given is to create teams, to choose team members and facilitators that stretch them into new paradigms, and to define team objectives without totally overwhelming your people. Do not shy away from this task; I can hear you pleading, "I am not an HR specialist!" You have risen through the ranks by getting things done, and that involves getting people to do things. You have knowledge, experience, and common sense. But I will give you some tricks of the trade.

Using Team Meetings to Proact

Team meetings are a necessity—not just for the sake of meeting, but to give team members a chance to get away from the normal reactive work environment in order to work on potential problems—so-called *proactive* thinking. I recommend one hour a week minimum away from the shop floor; if need be, teams can also conduct short daily meetings in their area to deal with day-to-day issues. Teams require time to think proactively in order to make improvements and solve problems.

Choosing a Team Facilitator

A facilitator does not function as a team leader or run the meeting; rather, the facilitator acts as a coach and coordinator. A good team facilitator is one who works through other people well; he or she makes team members comfortable and encourages them to get involved. A facilitator is not one who must control and have power over others, but he or she should be an excellent listener who helps people focus their thoughts and energy.

In early 1989, Valleylab hired a trained psychologist as manager of training and facilitation. He sat in on many meetings with the more than fifty teams throughout the company, working after the meetings with facilitators to review their techniques. This manager sees facilitators as "mirrors" of the team, helping team members to see how the team itself functions—a role he defines as "gatekeeper of the process."

The team facilitator is usually chosen by a manager or management team, according to his or her ability to help the team achieve its goals. Mature teams sometimes choose their own facilitators. In either case, the selection should be made carefully. The highest level person on the team should not necessarily be the facilitator; some very successful teams are facilitated by one of the workers while the supervisor is simply a team member. Sometimes teams depend so much on management that the team has no chance to mature on its own. In these cases, the area manager may be asked not to attend every meeting. An overly helpful manager might actually be perpetuating the team's dependence.

Managing Four Stages of Team Development

Teams typically evolve through four development stages. It is helpful to understand these stages so that they can be managed well and so that team members and their management won't get discouraged. Some stages seem unproductive, but so do a child's "terrible twos." But just as a child needs to attain that two-year-old's assertiveness, teams need to develop their own personality and ability.

Stage 1: Form

Forming a team entails assessing the type of team required and selecting members accordingly. Team members gain an understanding of the goal of the team and why they are being chosen for it. In this stage, team members should develop an understanding of the company's vision and how they fit into it. The team learns about JIT and about being a team, and the facilitator is chosen and trained.

Early on, the team develops a charter and objectives and determines the roles and responsibilities of its members. Ground rules for such matters as attendance, meeting times, decision-making methods, and conduct are decided upon. The members need time at this stage just to practice being a team.

One team member at AGFA Compugraphic describes this stage as "awareness, when you realize that there is a better way to do something."

Stage 2: Storm

As our Compugraphic team member puts it, this is the "awkwardness" stage: "This happens when you first try to do something that you are not used to doing. This is expected when trying new things."

Every team will go through the storming stage. Team members begin to take risks and challenge others, contributing and questioning things that have never been questioned before. Some teams pass through this stage quickly, while others need much storming before they are ready to go on. Although the stage appears to be full of sound and fury, it is not true that nothing is being accomplished; teams often make significant progress in this stage.

Being a good team does not mean that everyone is always agreeable and polite. Until feelings and perceptions are out on the table, they cannot be dealt with. When people are willing to put their feelings out—even if the ideas are unpopular—and deal with them openly, meetings are often stormy but productive. Until team members get honest with each other, they cannot progress.

Stage 3: Norm

Our AGFA Compugraphic authority calls this stage "skillfulness," saying, "This is the point where you feel comfortable with what you are doing, because you know that if it's done right it works." This stage takes a lot of thinking through and planning to get good results.

By this stage, people have become true team members. They have discovered how best to work with each other, utilizing their differences as team members to solve problems.

Stage 4: Perform

Our experienced team member calls this stage "Integration," remarking, "This is when it becomes second nature to you from continuous practice. It is a part of you."

Teams in the perform stage have achieved the cultural change that will last as long as the team is together. Being a part of a team is a way of life.

A team may backtrack into previous stages at any time as a result of membership or leadership changes or for other reasons. Team development is a continuous process. It is advisable to build team maintenance or "check-up" into the system.

The Five Essentials for Team Education

Teams move to the "norm" and "perform" stages much more quickly when enough time is devoted to education—not only as the teams are formed but throughout their development. Education should never be a one-shot process; it is a continuous process, providing what the team needs when it needs it. Since teams are made up of people and people are different, teams will have different requirements and progress at different speeds. Therefore, "generic" team education that does not take into account the makeup of the team and what its needs are will be unsuccessful.

However, there are five areas of education that should be addressed with all teams: (1) What is JIT? (2) What's in it for me? (3) Team training; (4) Facilitator training; and (5) Meeting management.

1. Teaching Teams About JIT

JIT education for teams covers the basics of JIT concepts and techniques from the perspective of the team. For instance, if I am educating a team that builds high-tech parts in a clean room under a microscope, I don't use examples involving building tables and chairs. If the team is from customer service, I don't use only examples from manufacturing. If the team is a part of a fabrication environment, with large equipment and many different products, I don't use examples and principles that apply to a repetitive environment, in which the same things are built over and over.

When planning how to approach this basic education with a particular team, think of looking through the eyes of the members and do what makes sense. I have found that the use of simulations (interactive, experiential education) is much more effective than classroom-style education.

2. Answering the "What's in It for Me?" Question

Part of what makes the ideas behind JIT click with people is relating it to *them*. Ideally, every team member will understand the impact these concepts and techniques will have on him or her personally, as well as on the department and the company, as "big-picture" company information is shared more than ever before.

Before you can realistically expect people to get involved in something new, you need to respond to the "me" question. People should be encouraged to ask this sort of question and bring up concerns relating to them personally. Validate, validate, validate! No question is selfish in this business. You can deal with these issues only if they are out in the open; if they exist and are not brought into the open, teams will be frozen.

3. Team Training

Team training deals with how to be an effective team instead of a group of individuals. Many of those interpersonal issues are addressed here. I reemphasize that successful handling of these areas is vital to get the team going.

Several basic skills stand out as essential for a group to become a team.

1. *Getting to know selves and each other.* For some teams, this is as basic as learning more about each other.

One team at Valleylab Stamford (a company owned by Valleylab) had thirteen members. Eleven were born outside the United States, in many different countries with different cultures and languages. One of the first steps was to get to know each other and to get used to communicating in a group. The group did a "life-line exercise" that allowed members to "draw their life" and then explain it. Each team member was given a piece of flipchart paper and a marker and was asked to draw his or her life without taking the marker off the page. Members included pictures of boats and planes connecting their homeland to where they are today. The team found a common bond in that almost everyone had made a big step when leaving a homeland.

Other methods of getting to know selves and each other include exercises such as risk-taking, goal-setting, and influence-styles activities.

I often do a Learning Styles Inventory* to find the different ways in which various team members learn. Some people need concrete examples; other people need to understand concepts first. Some need to do it themselves and literally "feel" the issue. The way people learn has a lot to do with the way they solve problems. Risk taking and goal setting also affect team development, as does influence style, the way in which different people seek to get others to see their points and follow their leads.

2. *Decision making.* How decisions get made in this new environment becomes a critical issue in every implementation that I have been involved with. It is important to emphasize that team, or group, decisions are not the only method to use in making business decisions. Authoritarian decision making, with one person making the decision, is still appropriate in some instances. Consultative decision making, in which one person makes a decision with input from others, can also be appropriate. The mode of decision making utilized within the business depends on the type of decision to be made.

There are several criteria to consider when choosing a method of decision making; they include time, quality, and buy-in. Figure 3-6 depicts the most common ratings for the different modes of decision making.

Teams need to determine how they will make decisions since they will

*Learning Skills Inventory is a self-scoring analysis created and sold by McBer and Company. It takes about ten minutes to fill out the twelve questions; then the person analyzes his or her responses, using a guide. When we give the LSI to a group, we usually spend a couple of hours going through it together.

Figure 3-6. Characteristics of different modes of decision making.

	Time	Quality	Buy-in
Authoritarian	Low	Low – High	Low
Consultative	Medium	Medium	Medium
Team/group	High	High	High

be empowered to identify and solve problems. Consensus decision making—making a decision that everyone "can live with"—is the preferred method for problem-solving teams. Team members are taught to listen to each idea carefully, asking questions for clarification. Each member should feel that his or her side has been heard. Eventually, a consensus is reached when everyone "can live with the decision"—it may not be everyone's first choice, but all members have been heard and can buy in to the team decision. They can support the team decision outside of the team.

The team is given an opportunity to practice the steps leading to a consensus decision with an interactive exercise. If members are just outvoted, they may stop participating.

3. *Problem solving.* A problem-solving methodology and various tools are required for problem-solving teams. A simple, easy-to-follow methodology or process for solving problems can help a team become functional very quickly. This includes training in getting to the basic cause of a problem, developing various solutions, prioritizing those ideas, and picking the right solution. A team can waste a lot of time and become very frustrated if some problem-solving structure is not followed. (Several problem-solving techniques are discussed in Chapter 4.) As with any skills training, the team will need time to practice each new technique through interactive exercises.

4. *Developing team charter, ground rules, and plan.* The steering committee, or the leadership driving the JIT implementation, should be clear and agree on the purpose of the team being formed. The goals set for the team should stretch it beyond the same old way of thinking. The role of the team and its members should be well thought out, with a clear scope identified—for instance, "from the delivery of material to the line until product is transferred to shipping." Some steering committees have used a "charge" statement to outline expectations for a new team.

The team then develops a charter, or mission statement, that clearly identifies its purpose. Of course, this "mission" supports the company mission for JIT and the overall implementation plan and challenge as determined by the steering committee. Having the team develop its own charter will test members' understanding of the team's purpose and force them to clarify in more detail the team's focus.

Next, team members should agree on how they want to operate as a team, a step referred to as setting procedures, or ground rules. Some common ground rules concern the following:

- Meeting frequency, time, and place
- Attendance
  —What constitutes a quorum?
  —Punctuality.
  —Priority with other things.
- Courtesy, such as
  —Don't interrupt.
  —Listen well.
  —Talk one at a time.
- Agenda, recording
  —How?
  —Who?
  —What?
  —When?
- Visitors policy
- Confidentiality issues

Ground rules should be reviewed and updated, if necessary, from time to time. It is important that the team be allowed to develop its own style and personality. There is no one way to have an effective, functional team.

The plan includes the team charter, along with goals, measurements, and milestones (the action plan). Writing a plan is the topic of Chapter 15, which focuses on implementation. Regardless of the makeup or type of team, ownership of the plan will be much greater if the team is involved in writing it. For instance, a steering committee made up of top management will have more ownership for implementation if it is involved in the plan development, rather than the top manager "passing it down."

The same is true at any level, with any team structure. It has always seemed natural to have managers write their own plans, but often managers assume that workers may not have enough planning skills to do a good job. Therefore, it has been common for management teams to set detailed goals for workers and then allow the workers to determine what to do to reach the goals. This may be necessary at first because of lack of experience and knowledge about what really can be. However, managers at Valleylab learned another lesson about workers: They plan very well with guidance and generally set more stringent goals than management would have set for them!

4. Facilitator Training

A team facilitator should attend all team-building sessions with the team. In addition, special facilitator training will provide the facilitator with skills in how to facilitate a group. Since each group is different, it is impossible to cover every challenge that a facilitator may face, but at least training can offer guidelines on how to handle common challenges.

Facilitator training covers working with the team in meetings as well as outside the team meeting.

Twelve Challenges Facilitators Face

The twelve most common challenges that facilitators face are:

1. How to get everyone involved.
2. How to prevent the extrovert from dominating without quelling his or her enthusiasm
3. How to get the introvert to talk more
4. How to handle the team member who is always negative
5. What to do if a team member doesn't want to be in the team
6. How to keep the team focused without directing
7. How to get everyone to volunteer
8. How to be one of the group and still facilitate
9. How to keep enthusiasm in the group
10. How to manage conflict within the group
11. How to get help when you are stuck
12. How to get resources when needed

How Facilitator Support Groups Can Help

Several companies have formed facilitator support groups as an opportunity for facilitators to get together and share frustrations, ideas, and successes, focusing on the process of facilitating a team. These groups have no projects; they are strictly support groups. I have found groups such as this to be one of the critical components for success in many companies.

5. Meeting Management

The way in which meetings are managed has a big impact on how successful the team is in accomplishing its goals. Again, team members should decide how they want to manage their meetings, but there are ten basic guidelines for making meetings more productive.

1. *Always use a flipchart.* Make magic markers and tape (to put flipchart paper on walls) standard tools for every meeting. If you write things so everyone can see them, the team will focus better and move along at a more efficient pace. Any team member can use the flipchart.
2. *Distribute prework in a timely manner.* The facilitator or scribe usually does this; but, again, any member can be involved.
3. *Set objectives for the meeting.* The team is responsible for this one.
4. *Agree on an agenda at the beginning of the meeting.* The team can generate the next agenda at the end of a meeting if desired, but in any case, give the team a chance to add/delete/change the agenda before beginning.
5. Prioritize the agenda items for discussion. The team is again responsible.
6. Estimate time needed for discussing each item. This doesn't mean that the team members can't change or extend times, but at least they have a guideline.
7. *Always* "process" the meeting at the end. Feedback at the end of every meeting on what went well and what needs improvement will keep the team on track. The facilitator typically leads this discussion.
8. Send notes, or minutes, out within a week. The scribe is responsible.
9. Seating arrangement can have a strong impact on communication. The facilitator should make sure everyone has eye contact with everyone else—don't allow members to sit off to the side. Try to have circular setups to promote good communication.
10. A team member who cannot attend should always agree to send a substitute. The facilitator should be notified in advance of a substitution or absence.

The Team Action Board

Teams at different companies have a variety of ways to manage setting agendas and priorities and gathering new issues and ideas. The best I have seen is AGFA Compugraphic's Team Action Board (Figure 3-7). It is not a new concept, but the simplicity of the application is almost startling.

Compugraphic uses the following guidelines and equipment: Large cork board surface; 3 × 5 index cards; thumbtacks; red, blue, and yellow markers; envelopes to hold cards.

Figure 3-7. The team action board.

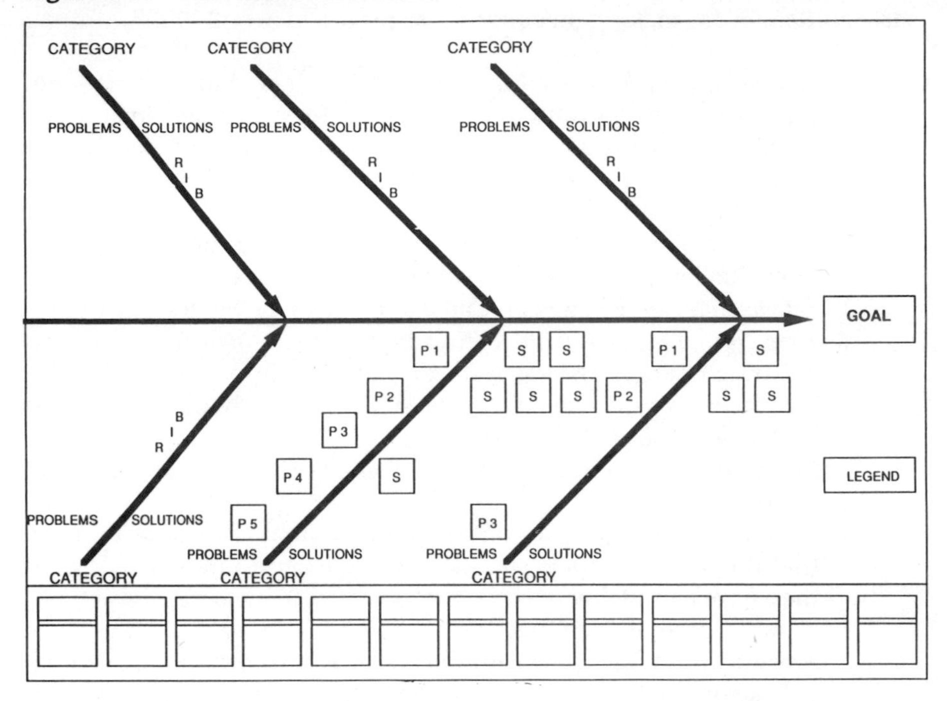

Board Setup:

- Create red ribs and spine. Leave space at bottom for envelopes, at rib extensions for category titles, and at right for goals.
- Teams should name ribs with appropriate categories for their area. Examples are: material, methods, machines.
- Issues cards are placed on the left side of appropriate rib. Closeness to spine indicates priority.
- Idea/solution cards are placed opposite the issue card on right side of rib.

Card Use:

- Employees record issues or solutions on 3 × 5 cards as they think of them and place cards on the Team Action Board. New issues are placed at rib ends; solution cards are placed opposite the issue.
- Issue cards are reviewed at team meetings. Once issues are accepted by the team, a priority for action is established and a blue indicator added to the upper left corner.
- Issues on which action is being taken are indicated by a red dot in the right corner.

- Solution cards reviewed by the team have a yellow dot in the left corner.

Card Filing:

- Closed issues are combined with solution cards and placed in an envelope. The envelope is marked with the work week, number of closed issues for that week, and the number of new issues for that week.
- The completed envelope is tacked to the board for the next thirteen weeks.

Team Action Board Benefits:

- Makes all ideas and issues visible to all interested parties.
- Assists in preliminary problem solving.
- Makes working issues with dates and names visible.
- Aids coordination of efforts by multiple teams that may overlap processes (such as first and second shift on same process).
- Creates secondary issue awareness (brainstorming) through visibility of issues.
- Eliminates need for paper lists or reporting on a regular basis.
- Facilitates meetings held at board.
- Enables peers to assist each other in making issues visible. People who are aware something is wrong and don't see a card up put one up.
- Allows ideas to be posted as thought of and not forgotten while members wait for a meeting.
- Adds consistency to visibility methods. Management can understand posted activity in all areas.
- Makes visible the team aspect of JIT to those who do not yet understand it.

Many companies have created their own version of the Team Action Board. It is a simple, visible way to manage as a team, regardless of the type of team or the level in the organization. One of the most significant benefits is that anyone can visit the team's area to see the accomplishments and challenges each team faces; this provides an opportunity to gather ideas from outside the team.

Three Aids to Keeping Team Momentum

Keeping team momentum is an ongoing issue in any JIT environment. After teams get through some initial stages and experience some of the easier successes, it is easy to fall into a lull—the newness is gone.

Figure 3-8. Sample self-critique form.

Stop	Start	Continue

I have found three aids that are very effective in maintaining enthusiasm and momentum: (1) self critique; (2) measurements; and (3) celebration.

Using *Self-Critique* to Keep a Team on Track

If a team takes the time to critique itself regularly, it never gets too far off track. This concept is similar to that of weight control—it is easier to take off those five extra pounds than to wait until the challenge is fifty pounds! Self-critique can include inviting an unbiased person into the team meeting to observe; on the other hand, the team can conduct a critique session on its own. One model that one of my colleagues taught me works particularly well: the Stop-Start-Continue model (Figure 3-8). Just make three columns on a flipchart and have the team list things it wants to stop doing, start doing, and continue doing. This will develop a good definition of what the team wants to be. On a more frequent basis, team members should feel free to ask, "How are we doing?" or "Are we where we want to be?"

Focusing the Team Through *Measurements*

Measurements against goals can be very effective in showing the team just how far it has come and in keeping the team focused. Without regular measurements, the team may not realize its accomplishments. I have found that measurements serve as a reinforcer much more than as a report card.

Using *Celebration* as a Reward

Celebrate those successes—regularly! Small celebrations validate success. Take time to reflect on what went right, so it can be repeated. We all need to feel completion as well as success to keep on going.

The challenge of developing effective teams throughout the organization is massive. It isn't easy, but once the new culture is in place, success is inevitable. Paying a great deal of attention to forming, educating, and facilitating teams early in the implementation process will pay off manyfold later on.

Chapter 4

Turning Quality Into Total Quality

Quality is one of the foundation blocks for implementing Just-In-Time. JIT cannot be successful without a prior or simultaneous commitment to quality. This is because JIT exposes problems and creates an urgency to solve them, while a quality effort provides the tools with which to solve these problems and maintain control.

It is possible to have a quality effort without JIT; most companies have problems that need to be solved permanently. However, JIT speeds the effort and assists in creating the right environment for quality improvement. In many ways, JIT and Quality are two sides of the same coin—you can't separate them, and one without the other isn't worth very much.

This chapter is not intended to be a technical explanation of quality and the many techniques, tools, and approaches for achieving quality—you can find volumes dedicated to these subjects. Instead, the focus of this discussion on quality is the mindset describing how JIT can help achieve quality and a simple methodology with which to begin addressing team problem solving in your organization. Some of the actions, as well as the methodology for problem solving, may seem just too simple, but I have found that with mastery of these things, any company will be well on its way to being world class.

First, it is important to outline four elements of a total quality mindset to ensure a common vision of TQ. Then I explain how JIT forces total quality and more about why the two efforts are inseparable. Finally, I detail eight actions to get going.

The Four Elements of a Total Quality Effort

"Quality" provides the tools for getting at the root cause of problems, solving them, and maintaining long-term process control. "Total" refers to the people side of quality that must be changed for true success with "Quality."

The four elements of a Total Quality effort are:

1. Meeting Customer Requirements

Meeting customer needs and exceeding their expectations 100 percent of the time is the ultimate success. Appropriate measures should be put in place to ensure customer satisfaction. Employees are driven to consider customer requirements first, then the company's requirements, then the team's—but always *all three*.

Quality should be positioned as a positive meaning "value to the customer," in the words of John Guaspari, author of *I Know It When I See It, The Customer Connection*, and *Theory WHY*.* Reminding people of what they feel is "value" as a customer will help in encouraging them to treat their customers the same way they like to be treated themselves. This brings quality down to a personal, individual level, which is where it all begins.

Not only should the external customer be considered but also the internal customer. Every action has a customer, and the goal of any activity is to meet that customer's requirement. The customer could be the next person in a manufacturing process or the next department that receives the work. If everyone were treated as a valued customer, how different the environment would be!

2. Building Quality in While Taking Cost Out

Manufacturers have always promised their customers high quality. To ensure that the customer received only good product, elaborate sorting and inspecting functions were created. The quality department was responsible for quality, while production workers were responsible for building as many goods as possible as fast as they could. Of course, cost went up as parts were scrapped or reworked and as sorting operations and inspectors were added. Either the customer paid more, profits were reduced, or market share was lost due to a price that was no longer competitive because of high quality control costs. Fuji Xerox Ltd. has shown that improvements in quality and cost can go hand in hand. Warranty costs in the field as well as reduced scrap and rework internally have been reduced. Management realized early on in this process that the cost of nonconformance was 20 percent of revenues.

Total Quality focuses on solving problems in the process *before* they

*All published by AMACOM in New York: *I Know It When I See It* in 1985, *The Customer Connection* in 1986, and *Theory Why* in 1988.

result in product that does not meet customer requirements. The focus changes from one of never *shipping* bad product to never *building* bad product. Quality can no longer be inspected in—it must be part of the process.

To ensure quality product, quality is required throughout the entire process, from suppliers to manufacturing to the customer. Customer repairs, sorting, rework, and scrap are all waste—they do not add value to the product. Therefore, they must be eliminated.

3. Creating 100 Percent Ownership for Quality

When I tour production facilities, I particularly enjoy talking to people, and one of the questions I love to ask is, "Who is responsible for quality here?" Many people will point to the inspector or the quality department and reply, "Inspection is over there." However, in a few companies the answer to that question is, "I am."

That is a big breakthrough. If I get that answer from the president, the president's managers, and even the janitor, then that company is really on to something! That is 100 percent quality owned by 100 percent of the organization. And that should be the goal in every company.

Ownership Includes Everybody

I will never forget the day I visited Digital Equipment Company's inner-city Boston facility. Digital built the plant hoping to employ inner-city people who have difficulty commuting to the suburban facilities of companies that had fueled the Boston area's high technology boom in the 1970s and early 1980s. DEC employed people from a number of different cultural backgrounds in this facility. In the beginning, this plant, which makes all keyboards for DEC, was struggling. The work force was unaccustomed to working in a high-tech environment. There were language problems.

Then the company formed a number of teams, focusing on quality improvement, and suddenly, the employees gained a lot of pride in their product.

A group of us came to tour the facility one day. We went into the company cafeteria to remove our coats prior to the tour. It was right after lunch, and a lady was busily mopping the floor. Suddenly, she put down her mop and came over to the group. She exclaimed that the plant built the very best keyboards in the industry, and unless we bought their product, we weren't getting the best! She went on to tell us how their quality is built in. Now that's really 100 percent!

4. Getting the Commitment for 100 Percent Quality

When I ask the group of manufacturing employees who would be happy with 99.9 percent quality, I always get a large show of hands. This is probably because 99.9 percent quality represents a vast improvement from the present quality levels in most manufacturing operations. But if our expectations of quality in all walks of life were 99.9 percent, the results would be:

- 20,000 wrong drug prescriptions per year
- More than 15,000 newborn babies dropped per year
- 2,000 lost articles of mail per hour
- Two short or two long landings at each major airport per day. (This one especially concerns me, since I fly at least twice a week for about forty weeks a year. I certainly hope the airlines have a goal of 100 percent good landings.)

Suddenly, 99.9 percent doesn't look so good.

If pharmacists, nurses, postal workers, and airline pilots can strive for 100 percent quality, manufacturing personnel can also.

Why would we ever ask for less than 100 percent quality?

You Get What You Expect

A company needed to order one hundred parts for a particular product. The company carefully packaged the order with specifications stating that a minimum 97 percent quality was required for acceptance—a fairly typical disclaimer statement. In a few weeks, the package of parts was received. When the box was unpacked, the receiving person noticed a small, separate package of parts—three parts, to be exact. Carefully enclosed was a note that said: "Enclosed separately are your three bad parts. We don't understand your request but are happy to accommodate your requirement."

The moral of the story, of course, is that you seldom get more than you require. People strive to *meet* expectations, not to *beat* them. Of course, a truly world-class company would ship a bad part only if requested to do so!

What if you, as the consumer, get the one of anything that is bad—the one wrong prescription, the one dropped baby, the one bad landing?

Recently, I was working with a team in customer service. The team members were very frustrated that manufacturing "never had the product when they promised." The 100 percent quality idea was unrealistic to this team. They felt it was better to have some product, even if a portion was bad. But that's just plain wrong. Bad quality can give a bad impression to the customer and have a negative effect on future sales.

Let's say that next week I hurry into the market to get eggs. The only carton of eggs has one broken. If desperate for eggs and short on time, I might take the carton, paying for twelve eggs but receiving only eleven good ones and a mess. However, the next time I need eggs, I will remember this experience and try the market across the street. As a consumer, I want consistent 100 percent quality and availability. One hundred percent is the only acceptable goal for quality.

You as a manager implementing JIT will find yourself facing the "100 percent requirement" sooner than you can imagine. I know some of these ideas sound unrealistic, but expectations on quality are rising very quickly. In the next section, I explain how JIT can help you accomplish this seemingly unattainable goal.

How JIT Forces Quality Improvement

Because Just-In-Time concepts and techniques focus on building only what is needed when it is needed, rather than building as many as possible as fast as possible, there should be enough time and flexibility to build only 100 percent quality product to satisfy customer needs.

Let's take a little closer look at just how JIT forces better quality. Remember the "I Love Lucy" show in which Lucy is packaging candy in a candy factory? As the candy comes down the conveyor, Lucy carefully places the candy in the box. Management decides Lucy can work faster and turns up the conveyor so it goes faster. As Lucy tries to speed up, she is forced to put the candy in her hat, her blouse, and in her mouth as she struggles to keep up with the belt. The quality of her work is disastrous because of the rate at which she is required to work.

The principle that says build-as-fast-as-you-can-and-don't-stop does not promote quality. Any mechanism that introduces parts to be built assumes each part will take exactly the same amount of time as every other part. If the time allotted is exactly the time it takes to assemble the pieces, people have no time to do an additional check or an extra activity to be sure of the quality of their work. There's another part to build right behind the last one. Automated conveyors can be one of your worst enemies in the 100 percent quality battle. Material moving devices force people to move on, whether they are ready or not. Instead of speeding up the work, these devices can actually slow it down by creating more scrap and rework.

I know this seems like a radical idea—automated material movement has been considered the way of the future. But please keep an open mind. (Automation will be discussed more fully in Chapter 6.)

Does this mean that the people don't care about quality? Of course not. The manufacturing system has set them up in such a way that they are not allowed to build 100 percent quality parts. Stopping is not acceptable in this environment; quantity is emphasized over quality. Creating a mindset that says it is better to go at a slower pace to build 100 percent quality and that it is okay to stop and solve problems forces better quality.

Figure 4-1 illustrates a production line for the production of widgets. In this example, operation 1 adds an A, operation 2 adds a B, and operation 3 is the final process that produces a widget.

I don't have to tell you that in the traditional manufacturing environment, operation 1 would build a batch of A's and pass the entire batch on to operation 2; operation 2 would build a batch of B's—working to build as fast and as many as possible—and pass the entire batch to operation 3; and finally, operation 3 would complete the batch and send it to inspection. As you know, large batches create a push to get all parts done fast. Less attention is paid to each part because of this push and because there are so many parts at once.

Eventually, the inspectors would inspect the product, and if they discovered that the entire batch had wrong B's, the wrong B's would be removed, the right ones added, and the batch of widgets sent on its way, with an appropriate note on the quality report.

The next week, during a department meeting, operator 2 would be told that one day last week, all the B's were wrong and asked to "watch the B's." Of course, he or she would agree.

In a JIT environment—with the rule that product is built only when the downstream operation needs it—operation 1 adds an A to one piece of product, then operation 2 adds a B to one piece of product, then operation 3 processes the product.

With time and ownership of quality, and only one piece of product to look at, operation 3 is more likely to notice that the B is incorrect. The product is then handed back to operation 2 for correction. Operation 2 stops long enough to resolve the problem; then it can resume adding B's to product as it moves through the process.

Yes, the line was stopped while operator 2 resolved the problem.

Figure 4-1. A typical production line.

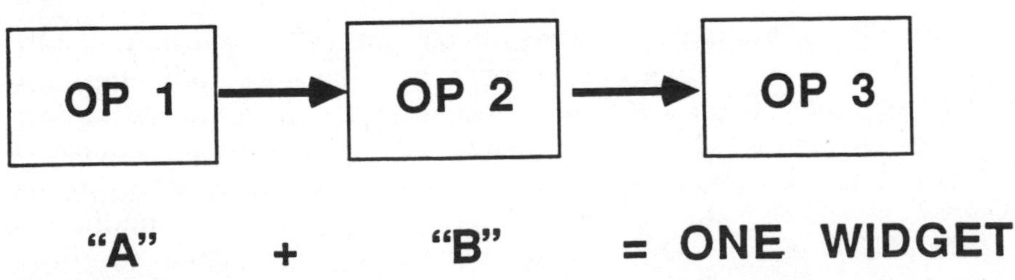

Operation 1 and 3 were idled also. But the alternative method caused an entire batch to be reworked. In the JIT setup, immediate focus placed on the problem thanks to timely feedback resulted in quick resolution of the problem.

In which scenario will operator 2 be more likely to watch the B's? I think you will agree that the second way is better.

Nine Major Actions That Ensure Total Quality

Although JIT concepts and techniques support the notion of taking enough time to get it right by allowing the line to stop and solve problems and by forcing less inventory and smaller batches, if there is no quality effort ongoing in a company—or at the very least being instituted simultaneously with the implementation of JIT—JIT will force a focus on quality.

Action 1: Know the Customer Requirements

Most companies assume that they understand customer requirements. In fact, frequently changes are made without consulting the customer because the supplier is certain that the customer will like the "new and improved" version better.

Ask, Don't Guess

Recently, a company made a change from a fabricated part that required special tooling and had some variability in quality to a purchased part that eliminated the tooling as well as the variability. What could possibly be a problem with this?

The company began shipping the product with the new purchased part, and the customer called, frantic: "I cannot use this new part—the adhesive that you used to attach the purchased part won't work in my equipment!" The company had never asked the customer or sent a prototype—it had thought that it understood the customer's requirements well.

This was an expensive lesson to learn, and the customer was hurt.

It is a good idea to do a customer-needs survey to make sure that the company understands the customer's needs. In some industries, this should be repeated on a regular basis, because things change rapidly. Requirements clearly stated from the customer can create an urgency to change within the organization. After all, who can argue with the customer? The vision, or focus, can be clarified through determining the true needs of the customer.

A survey to gather pertinent information can be done by a consultant or by your own organization. Many times, I do them jointly with clients. A written survey with phone or in-person follow-up is most complete, but even a phone questionnaire can help. Typically the following types of information are collected:

- *Background and factual*
  - Who is the respondent?
  - What are your products?
  - Details about your business?
- *Industry trends*
  - Equipment used and anticipated?
  - Budgeting process?
  - Criteria for upgrading processes?
  - Staffing?
  - Process Control?
- *Equipment use*
  - Types of equipment used?
  - Types of equipment for the future?
- *Buying process*
  - Lead times for purchases?
  - Who approves expenditures?
  - How are expenditures justified?
- *Satisfaction with manufacturers*
  - Overall satisfaction with certain suppliers?
- *Buying decision—usually rated and weighted*
  - Sales representative/customer service information?
  - Line of products complete?
  - New technology implemented?
  - Can provide technical help?
  - Financially stable?
  - Service and parts?
  - Meets delivery and price requirements?
  - Adds value?
  - Can respond to changes?

Another idea is to be sure that you and your management get out to visit customers regularly—set some goals and stick to them. One of the most potentially dangerous things for you to allow is to get so tied up with business internally that you forget about the customer.

Action 2: Help the Quality Department Become a Catalyst for Quality

The quality department must become a catalyst for quality, rather than merely an inspector of quality.

Typically, the quality department has acted as a "police" organization for sorting out problems. Members have had the authority in the past to decide if a part is good. Barriers have been strong between quality and manufacturing. Quality is considered a separate team, usually with its own area and a schedule different from that of manufacturing. The attitude of the quality department has typically been, "I'm going to *audit* these parts," implying, of course, that quality personnel expect to find bad ones.

It is easy to see how these barriers get built. Quality is the "police" and manufacturing, the "offenders." Manufacturing workers are evaluated on how many problems are found by quality. But constant worry about what quality will find can actually cause more quality problems than it solves. Here are three steps for breaking down the barrier between "quality" and "manufacturing":

1. Train quality people along with the manufacturing group. Make sure quality personnel understand their new role.
2. Make quality a part of the manufacturing team. Develop in-process inspection procedures and have quality move to the same area and hours as manufacturing. Quality personnel can help manufacturing personnel take responsibility for quality.
3. Train quality people in interpersonal skills. Their attitude toward manufacturing should be, "I'm going to verify your quality" or "Let's work together on this problem" or "I am here if you need me for consultation." These comments convey a "we're all in this together" attitude, implying they assume parts are good.

Now, this all sounds straightforward, but it is far from easy. These changes can cause major disruption in an organization and you as one manager obviously do not have the power to make all these changes unilaterally. So take it slowly.

However, if your company's top management is really serious about implementing JIT, leading this type of issue can help you get ahead of the game quickly.

Management in manufacturing and quality at Valleylab went through all the right steps to prepare for quality people to move out of their own area and onto the manufacturing floor where the final-assembly line for generators was located. However, emotions were high. The quality inspectors found many reasons why this arrangement would never work, citing the impossibility of testing in such a noisy environment and the problem of having the wrong type of benches. (Photo 4-1 shows a quality inspector in the separate area and all the product waiting to be inspected.)

It took several months of consultants and managers working with the manufacturing and quality teams, both separately and together, before they finally really worked on something as one large team.

It was the holiday season, and the production manager called people to come over immediately. I assumed that meant more problems. As we walked into the manufacturing area, we saw the entire group—quality and manufacturing—standing together and singing a song to a holiday tune with words about JIT. They had performed the song *together*. That was the beginning of something wonderful, and it has become better ever since. So has the quality of the product they build.

Photo 4-2 shows a quality inspector at her work station out in the manufacturing area—easily accessible to everyone and part of the flow. Notice the quantity of product waiting—a few product variations fitting on two shelves. (I talk more about what else happened to improve the material flow in Chapters 6, 7, and 8.)

4-1. Product piles up as quality inspector struggles to keep up in a traditional setting at Valleylab. Sometimes misunderstandings between manufacturing and quality impede efficient work flows; management must then step in and foster a sense of teamwork between the two.

4-2. A Valleylab quality inspector, working under JIT, checks out product at her work station on the manufacturing floor. Only a few items await inspection as a result of improved work flow.

Action 3: Train People to Know What Is Good

One of the greatest challenges in training people to know good product from bad is determining what really meets customer requirements. Everyone needs to work with the same clear specifications and procedures. Updates on customer needs should be complete and timely and accompanied by additional training.

Many times samples, such as a completed part for use as a model, help. Some companies are making the customer much more accessible to their people, providing direct communication that is accurate and timely. Figure 4-2a shows the traditional way that information is fed back from the customer. Every time the informatin is translated, the message may change and the time is increased.

In order to make communication more accurate and timely, direct lines of communication are required as illustrated in Figure 4-2b. Immediate feedback is one of the best training tools available.

Action 4: Develop a Consistent, Clear Approach for Problem Solving

If people working in the process are given a simple, straightforward methodology, the right tools to solve problems and monitor quality, and

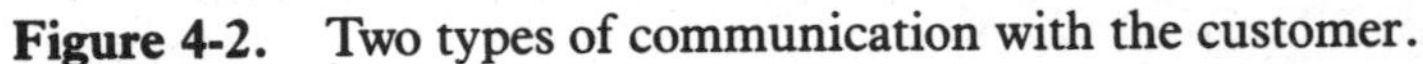

Figure 4-2. Two types of communication with the customer.

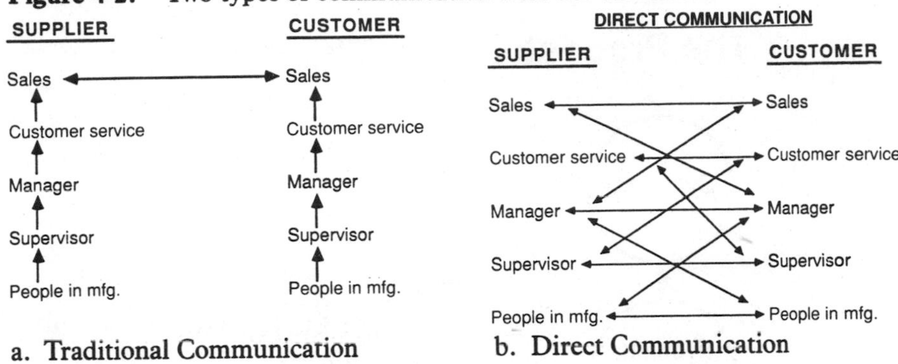

a. Traditional Communication

b. Direct Communication

training in how to use those tools, they can solve many of the problems in their work area. I find it essential that an organization develop a common methodology and vocabulary for problem solving.

A problem-solving methodology can be utilized to solve problems or to identify and implement proactive improvements. For instance, if the goal is to reduce cycle time by 80 percent, the same methodology can be used to identify the root cause of long cycle times and to develop a plan to attack that cause.

There are many valid problem-solving methodologies. The keys for success are that it be simple, clear, a linear process (step 1, 2, 3, etc.), and complete. We have developed a six-step process, outlined in Figure 4-3, that is easily followed. Team members can do quick checks to see if they are still "on track" with such a simple process; the steps can be posted in every team meeting area for clarity. Ultimately, it becomes a way of life.

Once a consistent, simple, problem-solving methodology is adopted, more complex approaches can be introduced as required.

Tools for identifying the problems include performing a value-added analysis (discussed in Chapter 1) and a flow analysis, which is a picture of the flow of the product through the process (see Chapter 6 for more information).

There are dozens of tools for diagnosing and solving problems. The eight discussed are especially successful and easy to learn to use.

1. *Five Whys:* A procedure through which you continue to ask why until you get to the root cause. Continuing to ask why may irritate some people, but it will drive people to think about possible causes.
2. *Brainstorming:* A tool used to think of every conceivable idea possible and then clarify each one. After a list of ideas is made, criteria for judging the ideas should be developed and each idea analyzed against the criteria for selection.

Figure 4-3. The problem-solving process.

The Process	The Steps
Symptom	
Problem Statement	**Step 1** Discuss and define the problem
Root Cause	**Step 2** Display the data "as is" and determine root cause
Best Solution	**Step 3** Generate possible solutions **Step 4** Select the best solution
Implement	**Step 5** Recommend, approve, implement
Measure	**Step 6** Measure, evaluate, adjust, and celebrate
Problem Solved	

3. *Pareto Analysis:* A statistical tool for getting at the significant things to work on. For instance, if your company were to develop five defect categories for analyzing problems and began to log problems into these categories, it would be possible to determine which defect category to work on first, according to the highest number of occurrences.
4. *Fishbone Diagram:* A tool for determining possible causes for problems or to categorize problems and develop solutions in an organized manner. An example of a type of fishbone diagram appears in Chapter 3 as the Team Action Board.
5. *Active Experimentation:* A method for selectively trying different ideas to learn more about the problem or its best solution in a methodical way.
6. *Force Field Analysis:* A tool used for getting at factors helping and hindering reaching the vision. If the hindering factors can be removed and the helping factors increased, then you can reach the vision.
7. *Stakeholder Analysis:* A way to help you think through who would be affected by the changes under consideration. This tool will assist in making sure everyone who might be impacted is included.
8. *Statistical Process Control:* A methodology that monitors the process to make sure that it stays in control by sending warnings when the process is headed out of control. The idea here is to predict problems before they cause bad quality. There are many more advanced tools and techniques associated with SPC that I will again leave for the volumes on technical quality.

Some of the common mistakes in problem solving are:

- Jumping from symptom to solution too quickly, before critical facts are gathered and the root cause discovered
- Not taking the time to collect and analyze data. Data "talk" when you give them a chance
- Letting opinions rule over fact
- Not testing and verifying probable causes
- Accepting problems as givens

Some of the common mistakes in implementing the solution are:

- Failing to involve and gain commitment of those who will implement or be affected by the solution
- Failing to cycle back through problem solving if unexpected problems occur

- Failing to develop complete action plans assigning clear responsibility for each task and developing standards and timetables
- Failing to follow up, monitor, and evaluate results

It has been said that a problem is not really solved until you can recreate it and then solve it again.

Action 5: Understand the Marketplace

Make sure everyone is aware of marketplace issues—what matters most in your industry to your customers. Competitive products should be familiar to everyone. This information is not just for managers and supervisors anymore.

Demonstrations should be held on an ongoing basis to show how your products are used in the marketplace. Awareness of the use of products creates a greater understanding of the importance of quality.

Competitive products can be displayed and discussed on a regular basis. There is nothing like a little competition to create more determination to get it right.

Action 6: Get 100 Percent Ownership and Commitment

In order to have Total Quality throughout the process instead of sorting at the end, everyone needs to take responsibility and become an inspector; in addition, the culture must allow people to talk with each other about problems they find in each other's work. It is not an easy task to get people beyond their natural defensiveness about criticism to the point where they feel the criticism is helpful and aimed at getting the process to work better rather than just a way to place blame.

Who Needs 100 Percent, Anyway?

Recently, I was in a client company that builds a component for the disc drive that Hewlett-Packard makes in Boise, Idaho, my former workplace. The operator was testing these parts as I was talking with him. He pointed to the rack of parts and exclaimed that he couldn't understand how HP could be so picky—they expect 100 percent quality all the time.

I told him that I came from HP in Boise and that I remembered the part. If the part was faulty, we had a disc crash that took hours to repair. If the customer received the drive and then the part failed, a customer's business could be brought to its knees, requiring manual recovery of data at great expense to the customer and a loss of credibility for HP. The operator had never thought of it that way.

Action 7: Allow Freedom to Change

Allow freedom to make changes. If people have the authority to determine that problems exist and, in fact, are held accountable for quality, then they have to be able to go the final mile and do something about quality problems.

These keys open the door to ownership and commitment for quality. Many solutions really are simple once people are empowered to do something about problems.

Empowerment Leads to Initiative

On the final-assembly line for generators at Valleylab, a problem with mixed-up wires was uncovered. Wires were color-coded and were required to be attached in a certain order. Assemblers formed teams and were given the authority to think of ways to solve common problems. As soon as the operator responsible was told that the wires were in the wrong order, she began to think of ways to prevent the problem from recurring. Soon, she had developed a tool for checking the wires before passing the assembly on to the next operator, a piece of cardboard with colored lines drawn on it served as her gauge. The problem went away.

In this new environment, the assembler did not wait for engineering to develop a tool or change the design or for management to decide what to do. She just went ahead and did something within her own power to solve the problem. The keys were empowerment and timely, direct feedback.

I believe that a major reason that we were able to accomplish so much at HP without examples and books and terms such as JIT was the fact that the HP culture allowed and encouraged change.

Action 8: Baseline and Measure Quality

As the manager, you can help workers and teams "baseline" their quality level and then measure how they are doing. Measurements may include percent defects, rework hours spent, and causes of defects sorted by number of occurrences or impact. Post the measurements where they are visible and make sure management pays attention. Charts posted on visible walls using flipchart paper or erasable boards work well. Then help the managers of that area with the spirit of, "How can I help you with this problem?" or "You are doing a great job—quality is improving!" will go a long way to reinforce the ownership that you are looking for. Quality charts can be maintained by the people, creating more ownership for their own quality.

Recently, I was working with a team to develop some quality measurements. The only quality measurement in the past had been the final quality inspection yield.

When I asked the final test person in manufacturing how many products she had tested today, she replied, "No one ever asked that before. They only want to know how many good ones I have." The new quality chart was simple: number tested and number good every day.

Measurements can be proactive in preventing problems as well. The idea is that the measurements often show a trend toward problems, allowing adjustments to the process *before* out-of-spec parts are produced. Statistical process control is a term often used today to refer to proactive measurement of processes.

Action 9: Implement Total Quality in Design

Quality Function Deployment (QFD) is a method used to facilitate the product design process, focusing on what is important to the customer. Products can be designed faster, with lower costs and higher customer acceptance, utilizing this method. Some companies refer to this as matrix product planning because it determines customer requirements and then translates them into language that each function with the organization can act on.

Design of Experiments (DOE) is one set of tools frequently used with QFD to test multiple factors at the same time. This makes it possible to design products that are easy to manufacture and that meet the customer's requirement even in less-than-perfect conditions.

Many quality problems occur because of complex designs that do not allow consistency in manufacturing. Products can be designed to be simpler with fewer parts. The Taguchi method, developed by the Japanese engineer, Dr. Genichi Taguchi, makes manufacturing processes less sensitive to disruption. For example, if the temperature in a plant cannot be stabilized, the Taguchi approach would be to develop a manufacturing process not affected by temperature.

I have mentioned only a few of the many methods and tools being used today by world-class companies for better product design. Again, there are volumes written on this subject, the intent here is to emphasize the importance of the change in mindset for product design to focus on the customer needs and expectations.

Implementing total quality in design is a longer-term process; all the more reason to *begin now!* (I talk more about design as a building block for JIT in Chapter 13.)

Like Just-In-Time, Total Quality is a mindset first and a set of methods, tools, and techniques second. If there is no mindset for 100 percent and if people do not answer the question, "Who is responsible for quality" with "I am," then the methods, tools, and techniques will be of little value and the quality effort will be short-lived at best. Top management needs to set the tone and send the message that 100 percent quality is the imperative. Determining ways to make everyone feel ownership and commitment for the quality function is key in any JIT implementation.

Chapter 5

Synchronizing With Customers While Using Capacity Evenly: The Drumbeat

Drumbeat is the third foundation block of the Just-In-Time pyramid. Planning of material and scheduling of production, capacity, inventory record accuracy, and bill of material accuracy are all of part of this building block. Although these business basics are not replaced by JIT, their management is simplified by implementation of JIT concepts and techniques.

A regular, linear schedule that requires everyone, from suppliers to the manufacturing plant to customers, to "march to the same, regular drumbeat" is necessary. The first step in achieving this Drumbeat must occur within the manufacturing plant. After you have proved that a product can be manufactured in a regular, linear way, then the Drumbeat can be extended back to suppliers and forward to customers. The ultimate goal is complete synchronization with the customer, utilizing capacity in an even and regular way.

You may be wondering how this is possible, since your customers do not always order product in a regular, linear way. The key here is always to drive planning and scheduling with customer requirements, not with some internal focus such as how to keep all people and machines busy all the time making product whether the customer needs it or not. Drumbeat works whether you are a repetitive manufacturer or a job shop that builds specially designed products to customer specifications.

Implementation of Drumbeat has to start somewhere, and usually, manufacturing is the catalyst. Cooperation among materials, manufacturing, and sales and marketing are critical to the success of Drumbeat. Support from other functions within the organization is required as problems are uncovered by manufacturing.

This chapter may seem theoretical at first, but Drumbeat is a key focus of the entire JIT implementation. If the emphasis is on the correct Drumbeat vision, quality, cycle time, and costs will improve. I therefore first explain this all-powerful term Drumbeat. Following this, I focus on the implementation of Drumbeat and the role of sales and marketing in creating a regular, linear schedule. (How and when suppliers get involved in Drumbeat will be discussed in Chapter 11.)

Meeting Customer Needs More Effectively With Drumbeat

Key words that describe a Drumbeat schedule might include:

Consistent
Predictable
Flexible

These may seem somewhat contradictory at first, but read on to discover how Drumbeat works.

The Remedy for EMP Syndrome

At Hewlett Packard, we had a strong case of what I call EMP (end-of-the-month push) syndrome, a common malady I still run into today in many companies. I was in the materials department at that time—materials was next to shipping.

A strange phenomenon occurred on a regular basis, toward the end of every month. Everyone in shipping began to move faster than usual, working more overtime, until dark circles appeared under everyone's eyes. It seemed that *everything* that would fit in a box was being shipped. Receiving people were all helping in shipping, the receiving docks were empty. I dared not walk through shipping lest I would get shipped.

Up in production, the lines were filled with material, and people there were also rushing and working overtime.

As soon as the first of the month rolled around, things changed. The receiving people were receiving material like crazy, but the shipping people were in recovery mode—moving at half speed. The production lines looked as if someone had come through with a parts vacuum cleaner—there were no parts anywhere to be seen.

Workers were not busy. When we had our regular plant meeting, shipping and production were congratulated for producing and shipping so much so quickly—we had made it once again.

The news wasn't all good, though. The toll that EMP took on everyone in the form of stress and frustration was showing up. Warranty costs were rising. The divisional management staff began to notice.

A new idea was developed. At the next plant meeting, the general manager announced that the first production line that produced the same amount of product each week of the month consistently the entire month and got it all shipped that week would win T-shirts and a trophy. Within a few months, one product line was consistently producing and shipping in a linear fashion. The team members proudly wore their T-shirts and displayed their trophy—the race was on. Before long, it was unacceptable not to achieve linear production.

How did the lines do it? Many things had to happen in purchasing, planning, production, and shipping. The reward system changed, and so did the performance criteria within the plant. This forced changes in many areas.

Using the Daily-Rate Mindset to Achieve Linear Operations

Getting to the daily-rate state of mind, in which the manufacturing plant employees are thinking of how much product is being made each day, regardless of the particular product mix, is a three-stage process. Some companies are driven by quarters today: Monthly is better than quarterly, and weekly is better than monthly. It may take some time and hard work to get to daily in some businesses.

Figure 5-1 shows a hypothetical master schedule in a manufacturing plant for a product line with variations, all requiring changeover of people, equipment, and/or tooling.

Figure 5-1. Traditional master schedule for widgets.

Options	Week 1	Week 2	Week 3	Week 4	Week 5	Week 6	Week 7	Week 8	Week 9	Week 10
A	27 →	→	→			17 →	→			
B			13 →	→			3			
C					10			30 →	→	→
Totals	27	0	13	0	10	17	3	30	0	0

Let's assume that the 27 A's scheduled for Week 1 are all sold. A customer decides in Week 2 that it must have an A quickly. Under the current schedule, the next time an A will be available is Week 6 at the earliest. This may not be good enough to keep that customer's business.

There is another problem with a shop floor that runs on such a schedule. Look at the bottom line—the total number of products to be started in the plant each week. The range is from zero to thirty. This causes havoc in purchasing, production control, stores operations, and production. Some weeks, there is not enough to do, and other weeks require overtime just to provide, inspect, store, kit and deliver parts, and manage the orders for the shop floor. In production, changeovers are many in some weeks and none in other weeks.

Stage 1: Smooth the amount of work done in each week to create a weekly rate. This uneven activity can be improved by making some changes internal to the manufacturing plant, such as reducing the order quantities and issuing new orders every week to support an equal start schedule, or bottom line. This results in a master schedule that looks like Figure 5-2.

But even in a plant with a schedule like that in Figure 5-2, the customer still must wait for the A's. Despite smoother plant operations, service is still not improved. Improving operations internal to the manufacturing operation is the first step in creating a steady Drumbeat.

By simply creating more orders, a Drumbeat of ten products per week can be created. This entails some extra planning, but can be accomplished, even if the products are not equal in capacity requirements, by producing a formula that creates an equivalency among the various products. Then a product mix can be developed to consume an equal amount of capacity each week.

Production control may have more orders to manage, but work is evened out in all areas. The plant floor benefits by having a smaller amount

Figure 5-2. Master schedule for widgets, stage 1.

Options	Week 1	Week 2	Week 3	Week 4	Week 5	Week 6	Week 7	Week 8	Week 9	Week 10
A	10	10	7			10	7			
B			3	10			3			
C					10			10	10	10
Totals	10	10	10	10	10	10	10	10	10	10

of in-process inventory on the floor. Each production area receives weekly deliveries of parts. If there is a parts quality problem, it is likely to be identified more quickly in this schedule.

However, the plant floor still has lumpy changeover activity. To alleviate this, a second change is necessary.

Stage 2: Alter the product mix so some of each product is made each week. Under the change illustrated in Figure 5-3, all products get built every week, so a customer requiring a product quickly has a better chance of getting it. By scheduling every product every week, flexibility is increased. If one less B is produced and one more A is needed, the changeover is already scheduled, so the impact on production is minimized. Regular changeovers and product build can stabilize the process, improving quality and productivity. In many environments, work orders are no longer needed.

Stage 3: Create a daily rate of work regardless of the mix. In this way, better customer service and a regular and even Drumbeat within the plant can be achieved. One of the keys to being able to produce every item every week, with an emphasis on the daily schedule, is reducing the time it takes to perform changeovers, the subject of Chapter 9.

The Four Benefits of Mixed-Model Scheduling

Mixed-model scheduling, or having the capacity to schedule the build of mixed models or products every day if needed, has four important benefits.

1. *Flexibility* is increased as the floor is able to build smaller batches more often as needed.
2. *Customer responsiveness* is improved as the manufacturing lead time

Figure 5-3. Master schedule for widgets, stage 2.

Options	Week 1	Week 2	Week 3	Week 4	Week 5	Week 6	Week 7	Week 8	Week 9	Week 10
A	4	5	4	5	4	5	4	5	4	4
B	2	1	2	1	2	1	2	1	2	2
C	4	4	4	4	4	4	4	4	4	4
Totals	10	10	10	10	10	10	10	10	10	10

is reduced through smaller batches. The right things are built at the right time.

3. *Marketing/sales strength* can be developed now that the floor has the ability to respond to customer needs.
4. *Better quality* is always a result of Drumbeat attainment and mixed-model scheduling. Quality improvements have been in the 50 percent to 100 percent range for companies who have achieved ownership of these goals at the worker level by implementing Drumbeat. As I emphasized in Chapter 4, haste makes waste. As schedules become more linear and predictable, workers have the time to build in quality. Also, as workers become synchronized with the customer needs, ownership for schedule and quality increase manyfold.

Three Ways Drumbeat Can Simplify Traditional Planning and Scheduling

If you are thinking two steps ahead of me, you have probably been asking yourself, "How does this all fit with MRP, MRPII, and CRP?"

Manufacturing Resource Planning (MRPII) is a procedure through which companies plan production and then execute the plan, feeding back information on how well the plan was executed. Material Requirements Planning (MRP) is a tool, usually computerized, to plan component part needs and to schedule manufacturing, so that the overall build can be executed. Capacity Requirements Planning (CRP) is a tool for planning detail capacity usage and needs.

A JIT Drumbeat in no way takes away the need to plan and schedule. Having effective planning and scheduling is important to the success of any company. But JIT simplifies many traditional planning and scheduling techniques. A few of the changes that may occur are:

1. Business planning and master scheduling (usually in product families) will have the daily-rate state of mind, forcing planning to be based on at least weekly time buckets, not monthly (traditionally, this has been the norm). Longer-term planning and scheduling become less detailed as flexibility to meet the customer's needs is increased.

2. While MRP may be required to plan components, more purchasing is done to support a Drumbeat. MRP is a batch system, and JIT is a flow system.

In other words, if the Drumbeat is to build ten units per week the purchasing department can simply ensure that it buys parts to support the ten per week. No complex system is needed to support a consistent build of product.

However, with complex products and job shop environments where products are designed and built to customer order and possibly specifications, a more complex system is required.

In either case, I have found that MRP for the planning and scheduling of components is required at some level of detail for every company that becomes world class.

3. Less detail is needed to schedule the plant, since flow lines that build a product, start to finish, replace disconnected work areas that each build a part of the product. Chapters 6 and 7 describe this idea of flow and some new techniques for scheduling the build of product in a flow environment. Therefore, the part of MRP that does detailed shop floor scheduling is replaced by the JIT Drumbeat.

How to Begin Implementing Drumbeat

The concept of Drumbeat makes sense; however, the challenge is how to begin such a massive mindset change for scheduling. It is possible to begin small with pilots and expand over time. You might pick one product line—possibly the most stable—and begin there. I have found that visible charting is the key to a successful Drumbeat implementation.

Creating Visible Charts for a Successful Implementation of Drumbeat

In one plant, management emphasized two goals—quality and linear production. In this case, daily schedules were introduced (product cycle times were short here).

Instead of thinking and planning in weeks, everyone began to think days. The daily schedule focused everyone on quick resolution of problems *today*. Figure 5-4 is an example of the chart (clearly visible in the area) that the people used to track daily schedule attainment.

As you can see, the production schedule of ten was made only on one day. However, this visible chart began to focus the team on attaining the daily schedule every day. If the schedule was missed one day, the goal was to recover quickly. In other words, if on Monday only eight were produced, then on Tuesday twelve were produced—they recovered quickly. But more than ten could be produced only if a line was recovering a missed schedule.

As is true in every case, all workers need to get involved in order to have determined drive to eliminate the problems that prevent schedule attainment. Everyone involved needs to resonate at a personal level to the customer needs.

First, tools are needed to highlight the problems; second, management

Figure 5-4. Daily schedule of 10 per day, stage 3.

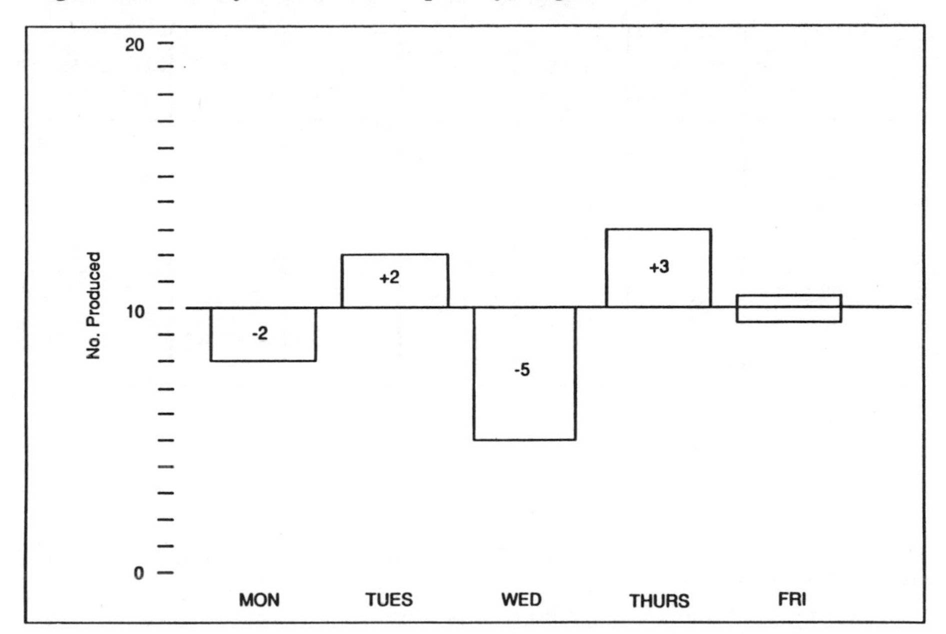

and technical support are needed to respond to these problems; and third, the appropriate people who can solve these problems should be empowered to do so. The Drumbeat chart is the most effective tool that I have found to highlight problems and measure the results of drumbeat (see Figure 5-5).

Guidelines for an Effective Drumbeat Chart

1. Make it visible in the production area.
2. Maintain the charts daily. All chart updates are best done by team members.
3. Log problems that prevent Drumbeat from being achieved. This provides a way for the team to make visible problems preventing its success.
4. Make sure that top management from every function is aware of and committed to Drumbeat charts. The team needs the support of management to help resolve the problems that are highlighted. This will require a walk around the floor *every* day by management from every function.

In some companies, what to list on the Drumbeat chart as customer needs is easy, as is how to align that need with capacity usage. A more

Figure 5-5. Drumbeat chart for widget team.

PART	MON P A Δ	TUES P A Δ	WED P A Δ	THURS P A Δ	FRI P A Δ	WEEK SUMMARY P A Δ
WIDGET A	1 0 -1	1 1 0	1 2 +1	1 1 0	1 0 -1	5 4 -1
WIDGET B	1 1 0	1 1 0	1 1 0	1 1 0	1 1 0	5 5 0
PROBLEMS	Part A Short		Overtime		Quality Problem Operation 2	

P = Plan
A = Actual
Δ = Variance

repetitive environment can usually synchronize capacity usage and "pieces" fairly easily.

However, in a job shop/build-to-customer-order environment, the task can be much more difficult. "Pieces" do not match up to capacity usage, and another unit of measure is required. Usually, these companies pick hours—either machine or people hours—as the standard. Items on the Drumbeat chart to be built may be customer orders and the daily (or weekly) goals may be "milestones" rather than completion of a product if product manufacturing lead times are long.

Each business will be different, but keys to look for include:

- Visible milestones no longer than one week that individuals can resonate to, reach for, and achieve (get the win!)
- Visible display of problems that prevented (or made it too difficult) to make the milestone
- Some method to ensure the right level of capacity usage

Red Flags to Consider Before Implementing Drumbeat Charts

1. Drumbeat charts cannot be considered report cards for the team. They are a tool for uncovering problems. After all, many of the problems that prevent schedule attainment are outside the direct control of the team.
2. Emphasis should be equal between *Drumbeat* and *Quality*. If ownership of Quality is forgotten, then you will be right back to "quantity, quantity, quantity." If the team places equal emphasis on Quality and Drumbeat, it will make the right decisions.

Therefore, the two essential visible charts for every production team are Drumbeat and Quality—both maintained by the team.

So, What Comes First: JIT or MRP?

The answer is: both. If your company already has an effective planning and scheduling procedure in place, then JIT will be very effective.

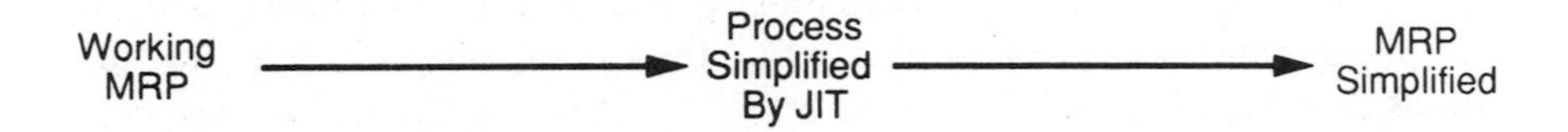

However, if your company does not have MRP, JIT can simplify and improve, eventually forcing the need to plan and schedule more effectively. Simplification, visibility, and process improvement forced by JIT will make MRP simpler and more straightforward.

A parallel implementation is also an option.

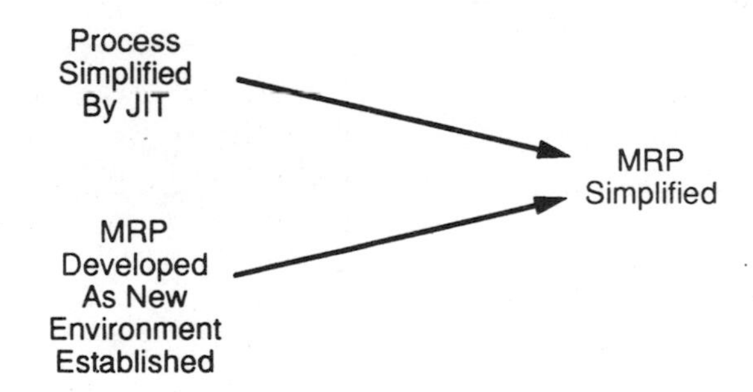

Systems will be discussed in more detail in Chapter 13.

Managing Changes in the Schedule

JIT creates more flexibility by increasing the ability to respond to change. However, that change must be managed. A Drumbeat will not succeed if there are constant short-term swings in the schedule. A regular, linear flow cannot be developed from a schedule that creates havoc. Therefore, schedule changes should be managed in such a way that they can be supported by all functions within the organization.

For example, Valleylab builds many standard variations of product. A particular product at Valleylab was forecast to sell as shown in Figure 5-6a.

However, as illustrated in Figure 5-6b actual sales did not develop as forecasted.

The traditional way to manage this discrepancy would be to adjust the manufacturing schedule quickly to meet the sales. But this causes havoc in the department providing purchased materials, as well as in manufacturing as shown in Figure 5-7.

How to Develop a Scheduling Model to Manage Change

Unless you are in charge of planning and scheduling, you may feel that you have very little influence as you seek the creation of new ways to plan and schedule. However, as data about Drumbeat are gathered and understood and the benefits of working on Drumbeat become apparent, you will get more support, one step at a time.

Top management support particularly is needed to create a model for planning and scheduling as part of the business policy for managing the company. This model includes the management of changes in the sched-

Figure 5-6. Forecast (left) and actual sales.

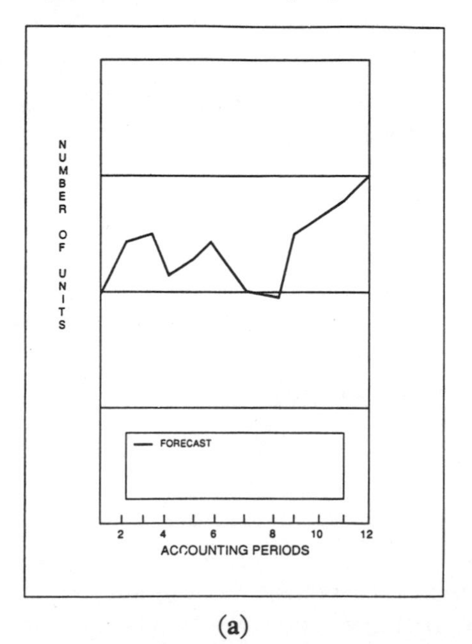

(a)

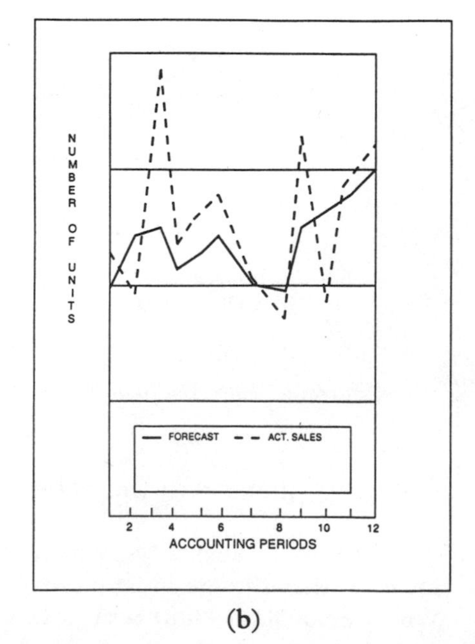

(b)

Figure 5-7. Manufacturing schedule.

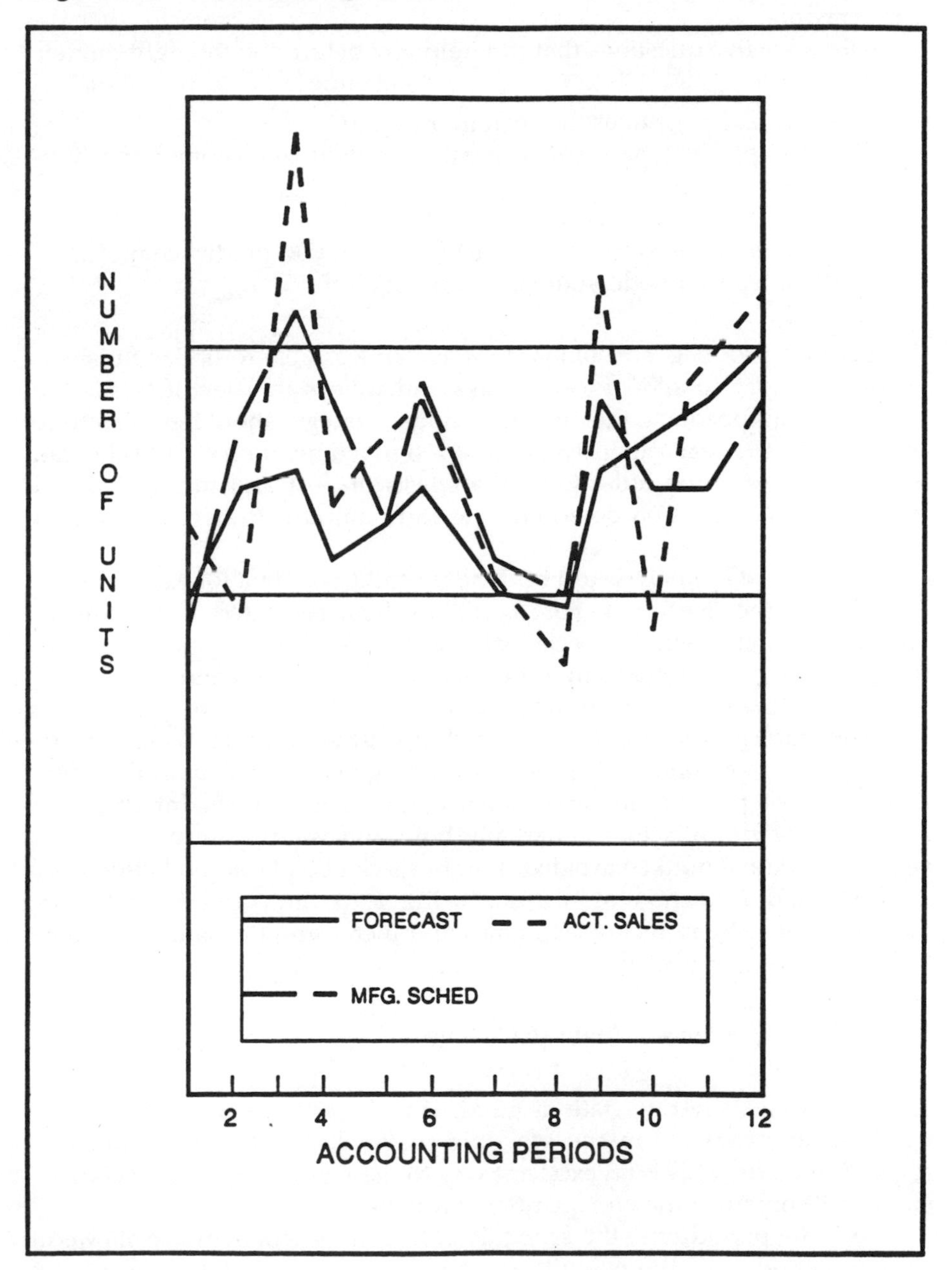

ule. Figure 5-8 illustrates the goal of this model: a regular, linear production schedule.

There are five questions that can help you determine the right model:

1. What lead time does the customer require?
2. Is the product standard or must it be built to customer specifications?
3. How accurate can sales forecasting be?
4. What is the product marketplace? How do you predict changes?
5. What lead times do your suppliers require?

Because products are standard, Valleylab was able to utilize finished-goods inventory to allow for the peaks and valleys in sales. If this is the case for your product, the finished-goods strategy should be carefully considered. It is very expensive to have too much inventory, but sales can be lost if there is too little. Careful analysis of past and future demand swings should help you determine the right amount to have in finished goods.

The necessary analysis can be done by gathering data for each product and plotting the sales in the past as well as those projected for the future. Future data may include special sales activities, discontinuation of one product line for another, or other marketplace information that may influence future demands for the product.

Sales forecasts should be measured against actual sales as an important operating parameter. If products are engineered and built to order, then it is not possible to carry finished goods inventory. Intermediate inventory, or product that is partially built and waiting for an order, is very costly and should be avoided. Emphasis can be placed on better sales forecasts and reduction of manufacturing lead time and flexibility in manufacturing. By reducing lead times and increasing flexibility, forecasting becomes easier and therefore more accurate.

Using Time Windows to Manage Change

To ease into a daily-rate state of mind, it is possible to set up windows within which changes in the schedule that will affect total output are allowed to occur. This is an excellent way to manage change. For instance, Figure 5-9 contains time and quantity windows.

For time period one, the schedule is frozen; no changes are allowed. In time period two, the quantity can vary plus or minus 10 percent. In other words, nine, ten or eleven units can be produced, although ten are planned. The windows can open or close with regard to either time or

Figure 5-8. A model for smoothing the schedule.

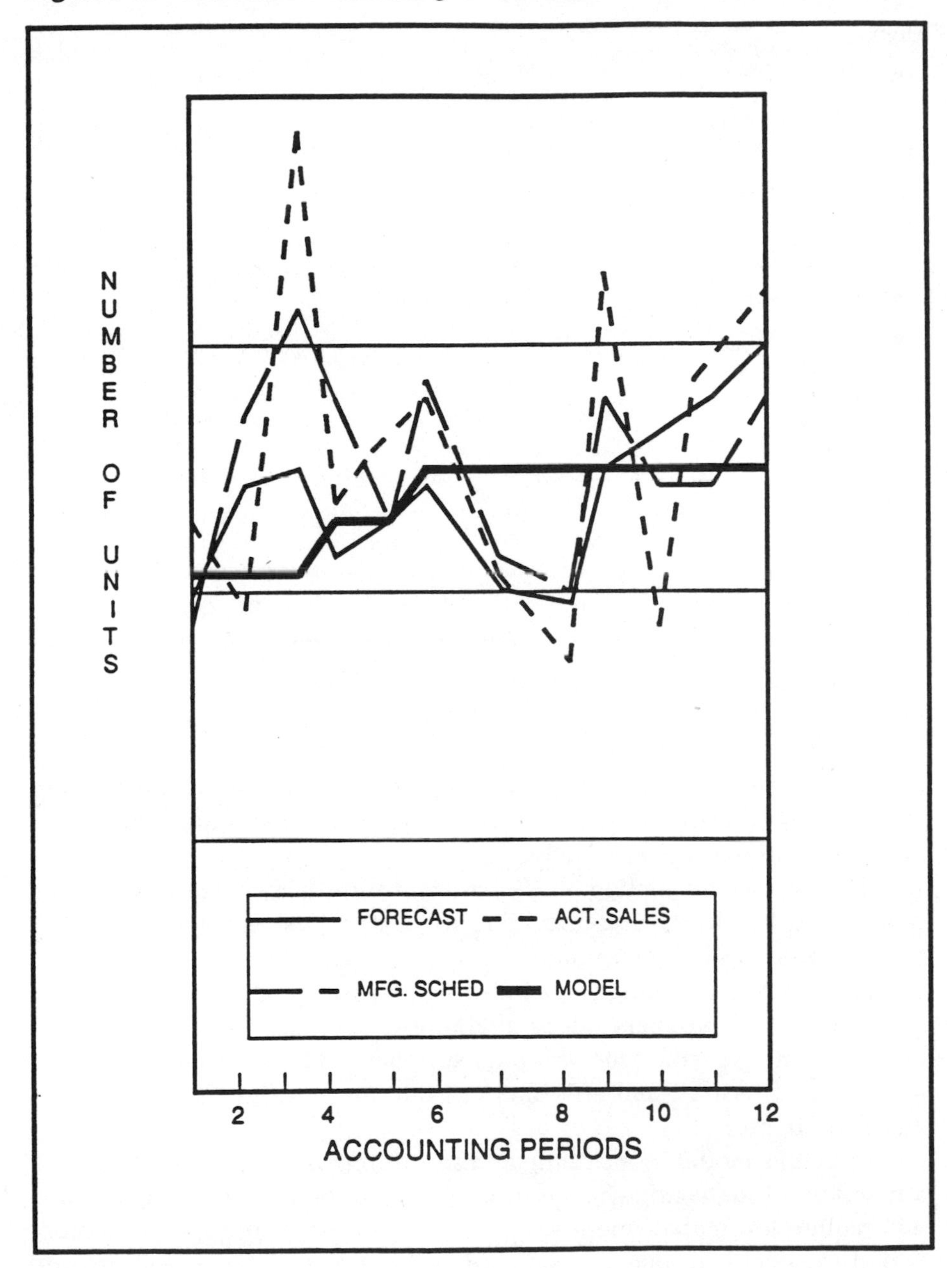

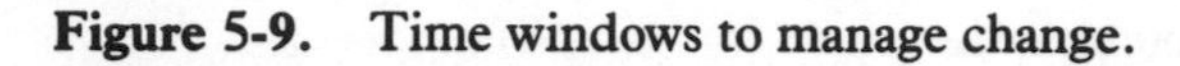

Figure 5-9. Time windows to manage change.

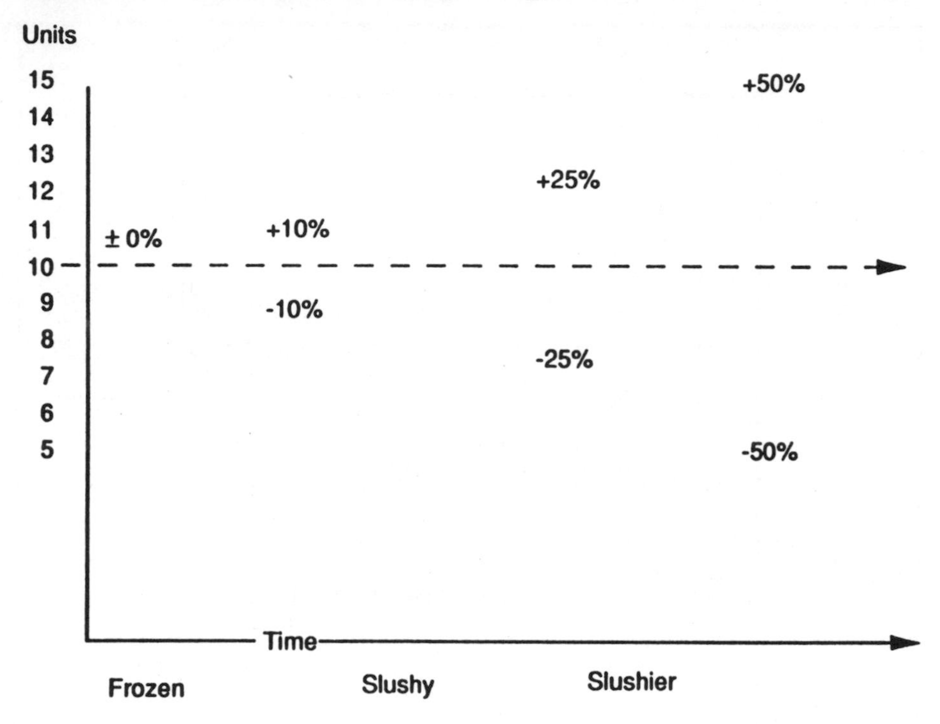

quantity as needed. But the numbers must be firmed again before moving into the next frozen period. Therefore, flexibility for schedule changes is not stifled but managed carefully.

The criteria for setting these variability and time windows can be determined by clearly understanding the entire business and answering the five questions for determining the right model. Successful development of this type of scheduling model also helps manage changes in supplier requirements. If suppliers can be guaranteed purchases of a certain quantity range in a specific time window, supplier lead times can be reduced significantly, creating more flexibility. I talk more about the role of suppliers in Chapter 11.

Once this model for planning and scheduling is developed, it becomes a new way of managing change throughout the organization. Again, this will require top management support. The important thing to remember is that the goal is to manage change in such a way that you will provide outstanding service to your customer—never to stifle change or customer service.

How Sales and Marketing Can Benefit From, and Help Achieve, the Drumbeat

Sales and marketing can help achieve the Drumbeat. Typically, sales and marketing are not involved in a JIT effort. Furthermore, when they hear the term they immediately hate it. After all, JIT sounds as if it means you will get parts built just in time to ship to the customer and if anything goes wrong (and it always does), the shipment will be late and the customer will be unhappy.

It is important that you help sales and marketing personnel understand the real meaning of JIT by getting them involved up front. They can benefit not only from the results of better customer service but also by applying the concepts and techniques to their own process.

Sales and marketing can actually use JIT capabilities as a sales and marketing tool. By being able to offer better deliveries, more often, with 100 percent quality and flexibility to respond to customer needs, many companies have increased their market share. But sales and marketing people cannot sell these benefits if they do not understand and believe in them.

The sales and marketing force can also help you achieve a more linear schedule by asking the customer to accept smaller shipments more often. In other words, if you ship a quarterly quantity to your customer today, ask about monthly shipments. Explain to the customer that this effort will help you provide better service by allowing more timely feedback and easier flexibility if requirements change.

Do not forget sales and marketing people when embarking on a JIT effort. You will find that their understanding and involvement can be critical for success.

Developing a Drumbeat mindset is a critical step in creating the successful JIT environment. That daily-rate state of mind supports the continuous improvement attitude that encourages finding the little things that can be made better today.

Chapter 6

Creating an Even, Balanced, and Self-Adjusting Flow of Product in Manufacturing

Now that I have discussed the foundation for a successful Just-In-Time implementation, it seems appropriate to talk about the "inside-the-factory" issues companies face in getting their own house in order. Typically, it is necessary to concentrate on these issues before tackling the "outside-the-factory" items. The exception is purchasing, which should begin pilots very early in the implementation. The order in which these inside-the-factory building blocks are assembled depends on the particular issues your company faces.

The first issue is flow, or how product moves through the manufacturing process. You can think of product moving through the manufacturing line as water moving down a stream. In a stream, water flows at some pace, self-adjusting for rocks and other obstructions; "bunches" of water do not remain in one place. Sometimes the banks are wide and other times narrow, but regardless of variabilities in volume the river flows constantly.

In manufacturing, the obstructions are all the problems that get in the way of building good product, and the banks represent capacity at any given time. The goal is to create and focus on an even, balanced, and self-adjusting flow of product in manufacturing, even if the volume varies according to customer demand.

In this chapter, I discuss how to document and critique your product flow, explain tried and true implementation steps for improvement of the flow of product, and describe how automation impacts flow. First let's establish what I mean by process flow.

How a Better Product Flow Can Decrease Cycle Time and Improve Quality

Figure 6-1 is not a spider web but a product flow drawn by a manufacturing team at Hewlett-Packard. When the team began to document its product flow, it tried to make a road map of where the product traveled to get built. Team members thought this task would be straightforward and simple. How they were surprised! They had a difficult time even agreeing on the process. It turned out that what the team really did to make the process work differed from the written documentation. Over time, the team had developed its own procedures to work around problems.

Finally, the team agreed that this flow diagram was correct. One important item to note is that the team—not the engineer or the manager—documented the flow. The team building the product knows the *true* flow better than anyone else. Let's take a closer look at this process.

Notice the intersection at A. It reminds me of a Boston intersection. When I first came to Boston, I was driving with a friend who was from the area. I came to an intersection where five roads suddenly converged—no traffic light, stop sign, or other indication of who should go when. I patiently waited for someone to motion that it was now my turn. Finally, my friend remarked that if I didn't just go, I would sit there until I ran out of gas! The rule in Boston is: Look straight ahead and just go. This can be dangerous, but it is the only way to get through the intersection.

Back to the example and intersection A. What all too often happens if your product flow includes an intersection such as this is that the products crash, creating scrap and rework. Parts go the wrong way, creating quality problems. And parts get lost. Losing parts seems to be a problem in every manufacturing environment. Sometimes I think that all manufacturing facilities have floors that open and eat parts!

Look at the distance the product travels—up and down and side to side. Again, parts get lost, damaged, reworked, and scrapped more often as the mileage increases. In addition, it takes much longer to travel further, and the need to count, track, and move parts increases with additional mileage. Waste is rampant in this process.

Notice the diagonal line labeled B. This is a typical detour—look at how it crosses several other paths. Again, it reminds me of driving in Boston. I have decided that the road crews tried to save money by purchasing only one-half of the detour signs really required. Every time that I take a detour, I get lost! The same is true of our parts trying to travel through this product flow. There are no specific queues in Figure 6-1. The rule here is: Store it wherever you can, but just keep building.

Figure 6-2 is an improved flow for this product. As you follow the flow of the product, notice the queue squares—four in some locations, eight in

Figure 6-1. Original product flow.

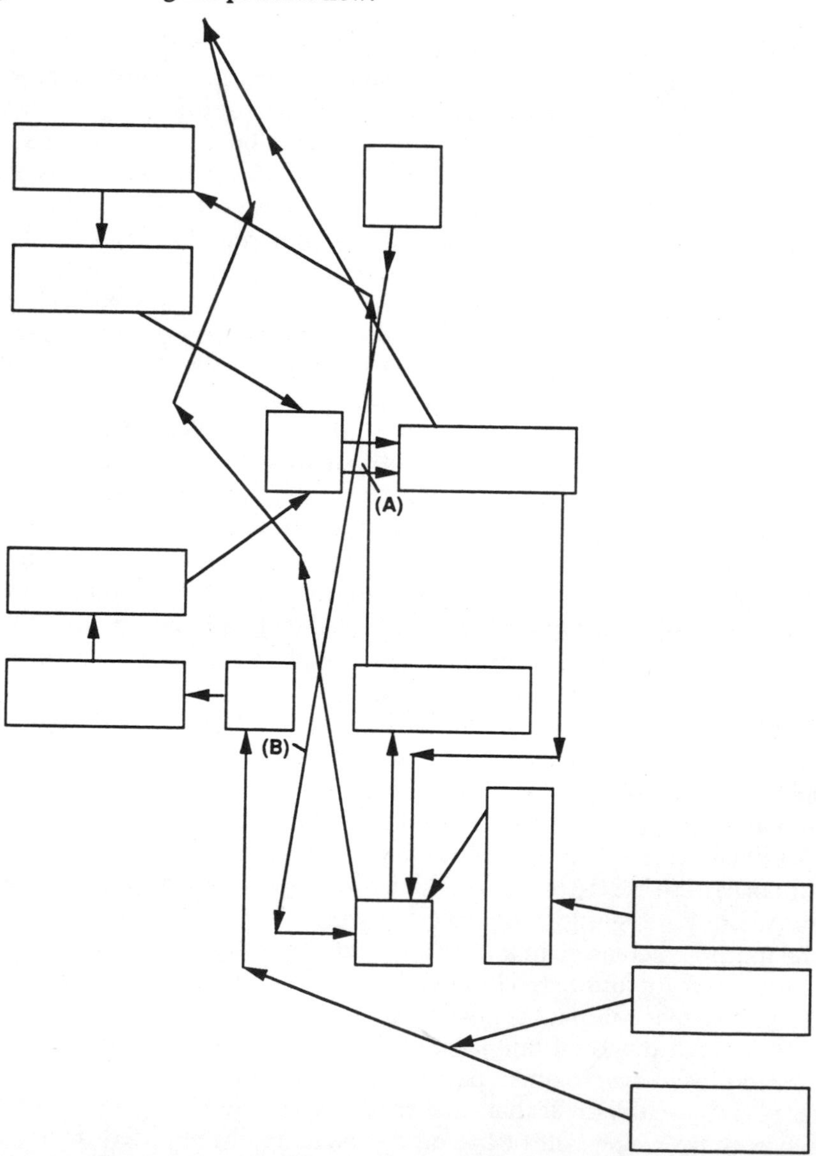

Figure 6-2. Improved product flow.

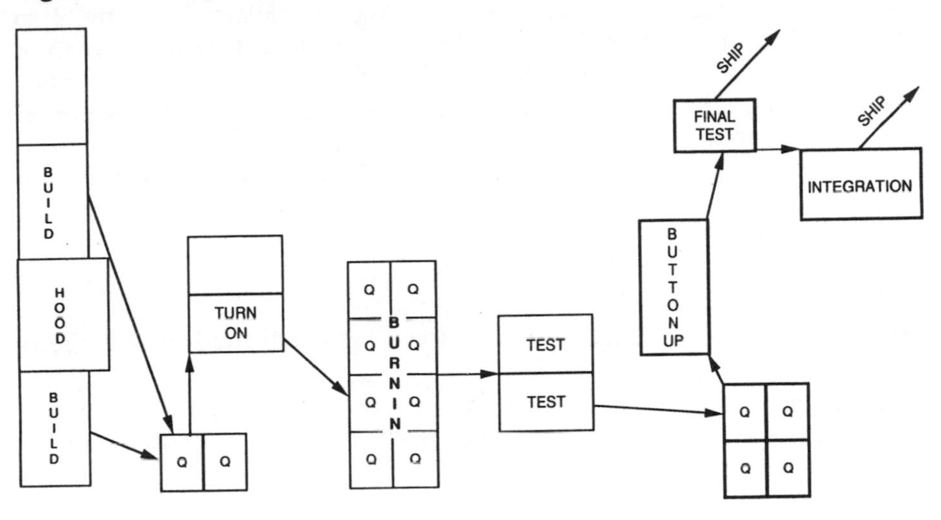

others. There must have been specific reasons for these different queue sizes. In any case, the flow is very easy to follow.

As the flow of the product improves, the cycle time, or elapsed time from the beginning of the process to completion, is reduced. Shorter cycle time, or speed, has become a major competitive weapon. Companies with shorter cycle times benefit in increased market share, customer satisfaction, and profitability.

As we have seen, quality also benefits as the product flow is improved. Haste does not make waste. An improved product flow can lead to significant benefits in customer satisfaction. The process of improving the flow can happen quickly, giving teams an exciting project that has tangible benefits.

Documenting the Product Flow to Discover Improvements

There are three key rules for successfully documenting product flow:

1. Always have the employees document their own process. Engineers and production management should provide only support, guidance, and information.
2. Document what *really* happens in the process—not what the procedures say.

3. Train the team before beginning on exactly what product flow means and how to document it. You can relate this activity to having a topographical map. The map represents the floor plan in manufacturing. Then add roads to the floor plan as parts travel through the process. Usually, just drawing the product flow generates many excellent ideas for improvement. Most people in manufacturing don't realize just how bad the flow is until they see it on paper! Once the product flow is documented, it can be critiqued and improved.

Six Criteria to Use in Evaluating a Product Flow for Improvement

Product flow should be evaluated and improved by using the six criteria discussed below. You may have other product-specific criteria to add to this checklist.

Criterion 1: Material Is Always Moving to Ensure a Constant Flow

The product should always be moving instead of waiting. Stand back and watch the process—how much is material moving instead of sitting in a bunch somewhere? The ideal is for material to flow through the process start to finish without any "bunches."

In the examples from Hewlett-Packard, more undefined queues are required in Figure 6-1 because of the confusing flow and batch requirements.

Criterion 2: Product Is Moving the Shortest Distance to Avoid Wasteful Mileage

The best flow is the shortest distance. This is a good measurement that can be easily quantified—just measure the mileage that the product travels. Figure 6-1 shows significantly more mileage than the improved Figure 6-2. Chances of loss, damage, mix-ups, and wasted time decrease as the mileage decreases. A focus on decreasing mileage by the team will also result in a clearer product flow.

Criterion 3: The Flow Is Well Defined to Prevent Confusion

Quality problems are inevitable when the flow is not well-defined. Confusion causes wasted time as well. The ultimate test of a clear and visible definition of the flow is to bring someone unfamiliar with the process into the area and ask that person to determine the flow by watching, with no help from flow diagrams or the people familiar with the process. This could be done in Figure 6-2, but certainly not in Figure 6-1.

Criterion 4: Operations Are Linked to Eliminate Choppy Flow

A client company made three subassemblies in one process area. These three subassemblies eventually became one product, as illustrated in Figure 6-3. As a batch of each subassembly was built, it was put into the storeroom to "maintain control." Subassembly A was eventually used for subassembly B, and then B was used for C to build the product.

As the team began reviewing the previous steps, it shortened the lines by taking out some of the work-in-process racks and stopped sending parts in and out of the storeroom. This improved the product flow mileage and eliminated the waste of going in and out of the storeroom, but the flow was still choppy (see Figure 6-4).

The next step for this team was to link the processes. Once the "straight line, one-way flow" syndrome was broken, it was easy. Figure 6-5 shows that only about one-half the previous distance is now traveled, and the processes are linked.

Criterion 5: The Product Flow Supports a Regular, Linear Flow—The Drumbeat

If the process is laid out to create an even, balanced, and self-adjusting flow, the Drumbeat will be achieved more easily. Problems are visible and can get focus quickly. Product flow through the process is predictable, and

Figure 6-3. Choppy flow.

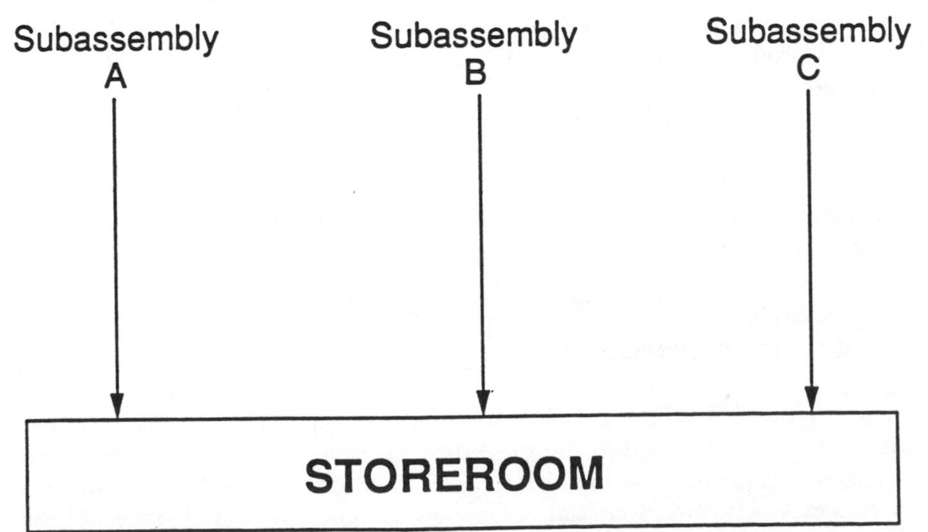

Figure 6-4. Linked operations.

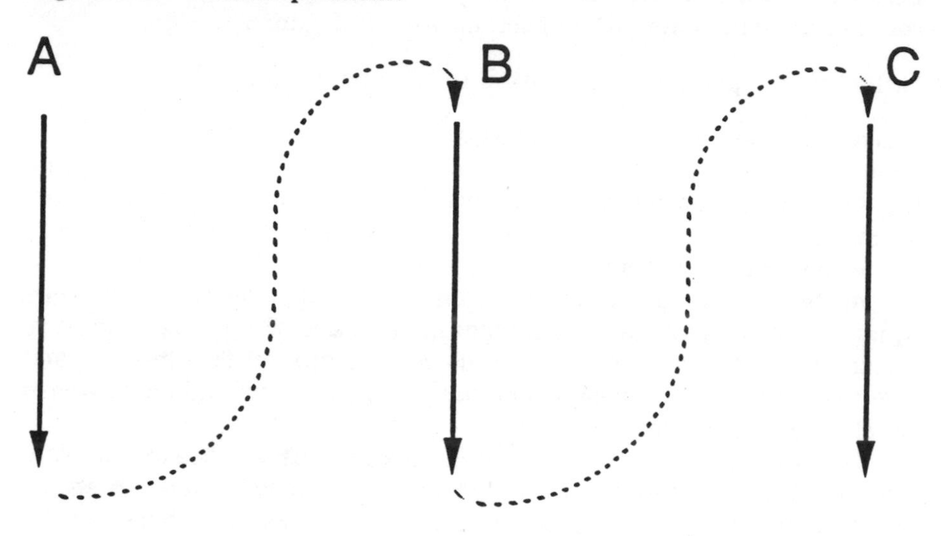

a regular, linear schedule can be achieved. The process depicted in Figure 6-1 is difficult to predict. It offers many opportunities to get off track, with dangerous intersections, extra mileage, and detours. However, in Figure 6-2, the flow is visible and predictable. Queues are planned and specific. There is more opportunity to focus on reducing these queues over time as the flow improves.

Criterion 6: The Flow Allows Maximum Flexibility to Support Changes

A good product flow assumes that changes will occur and adjusts easily for them. People, machines, layout, and tools should have maximum flexibility to allow for problems, capacity constraints, and swings in demand. You can make it easy for the team to make changes in its process by providing flexible resources such as power and air. If you have the opportunity, think about providing these types of resources on a grid, rather than in straight lines, as most plants are designed today.

Six Steps to Implement Product Flow Successfully

Once the present product flow is documented and evaluated according to the six criteria discussed in the preceding section, it is time to think about changes that make sense. The process used to develop and make changes in this new culture is critical. Employee involvement may take a little

Figure 6-5. Operations linked to minimize distance.

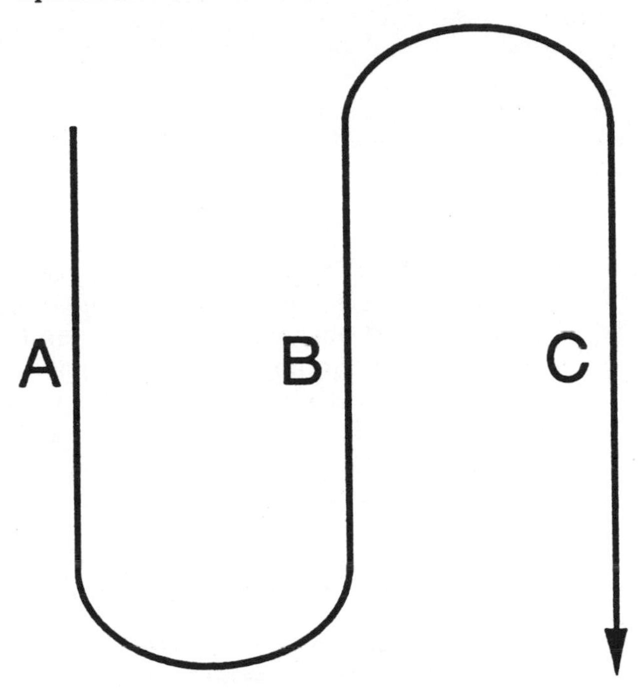

longer the first time than if only the engineer and production manager or supervisor were developing the changes, but the payoff in the long run will be far greater if you consider these helpful hints.

Step 1: Think Flow, Not Batch

A flow mindset can be created regardless of product or process. A company that engineers and manufacturers metal buildings to customer order would be a "job shop" in most of our minds. A job shop typically builds only to customer order, usually in batches, and sometimes even designs products to customer specifications.

But the president of a small modular building company that does this decided that he really has a flow shop: After all, all buildings have walls, doors, windows, and other common features. Even though each individual building may vary, the components are similar. And, since every one has a design phase, design is included as part of the flow. Today, each building that is started flows through the process as a part of the Drumbeat.

Step 2: Organize the Workplace for Quick Wins

Problems cannot begin to be discovered and solved until basic housekeeping has been taken care of. Organizing the workplace can often be undertaken while teams are still in training; it usually requires no management approval and can be done with little or no cost. Some key concepts of this organization are:

- Housekeeping with a purpose
- An assigned place for everything
- Cleanliness
- 100 percent participation

"Have a place for everything and everything in its place." That's the watchword here. It sounds so simple, but many people forget about it most of the time. Facilities are cleaned right before the president tours or on a special day when there is nothing being built, not as part of the everyday routine. But general housekeeping has a lot to do with how people feel about their company. If the place is dirty and in disarray, workers' pride is likely to be low. However, if everything is clean and neat—not necessarily fancy—there is pride in the work place.

Team Involvement

A team recently made housekeeping one of its goals. Members decided to have a ten-minute audit every day for housekeeping. This had not been part of their daily lifestyle, so they decided it would be necessary to enforce a particular time to pay attention to housekeeping until it did become a part of their life naturally—behavior modification in its simplest sense.

A manufacturing floor is not the easiest environment to keep clean and neat, but it can be done. I have seen metalworking facilities without a scrap of metal on the floor because the operator regularly sweeps the scrap off the floor. The operator also cleans the machines as part of setup. I have seen a rubber molding facility without a scrap anywhere. Someone comes by regularly and picks up the scrap that has been carefully placed in a bin by each operator as part of his or her job.

Manufacturing plants tend to have concrete walls and metal beams. One facility I visited simply bought some bright paint and let the people design their own walls, creating a bright and cheery environment just for the cost of some paint.

Good Housekeeping

One team from an electronics company was responsible for the final assembly of the product. As the team began to analyze its product flow, members noticed that all three operators spent a lot of time just looking for the proper tools to assemble the product. The tools were small, and operators tended to borrow tools from each other to get the job done. One team member remarked that he thought he could solve this problem if he could get $5 from petty cash and make a trip to the local hardware store. The supervisor of the area handed him the $5 and said, "See you later."

He returned with three cans of spray paint, each a different color. Workers were asked to stop what they were doing and sort tools so that the three operators had only their own tools at their tool cabinet. Operator 1 had all tool handles painted red, with the top of the tool cabinet also red. Operator 2 got blue, and operator 3 used green. The problem was solved. Even if tools were borrowed, it was easy to remember to return them, since it was so obvious that they didn't belong at the borrower's work station.

The idea of 100 percent participation emphasizes that everyone should "own" housekeeping for his or her area. At HP, the person in charge of the copy machines must have gotten tired of picking up after everyone else, because one day there was a new sign over every copier that read: "Your mother doesn't work here!"

Step 3: Consider New Layout Principles That Promote Flow, Problem Solving, and Flexibility

What's Wrong With Straight Lines?

In the past, production lines and administrative areas have been laid out to discourage people from wasting time talking. A typical layout included straight lines, which seemed like the best way to get a one-way flow, with work stations hanging off the line and work-in-process inventory placed between workers. Conveyors, motorized and manual, were common as the basis for these straight lines. Figure 6-6 is typical of the layout taught in most industrial engineering classes even today.

But this kind of layout greatly inhibits flexibility. If operator 2 is

Figure 6-6. The straight-line production layout.

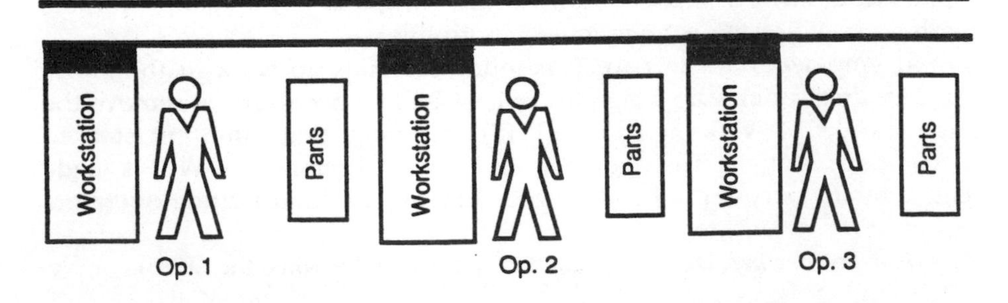

absent, then operator 1 or 3 must get up and walk around to fill in at work station 2. It is impossible to see work station 2 from their stations to know exactly when work needs to be done. Piles of parts make it difficult to tell what is needed and when. Certainly operators cannot talk without leaving their station. Problems can become overwhelming before an operator is willing to stop working and leave the work station to get resolution.

A good example of this type of layout existed in Valleylab's disposables area. It takes two photos (Photos 6-1 and 6-2) to show the entire length of the line!

How U-Shaped Lines Promote Communication and Flexibility

What we now know is that the best way to encourage problem solving is to get people to talk to one another. Better communication leads to better quality and schedule attainment.

In Figure 6-7, the line has been wrapped into a U shape. U-shaped lines are a classic JIT layout. However, the line can be any shape—even if it doesn't form a letter of the alphabet—so long as people are arranged in such a way that they can easily communicate about the process—both improvements and problems—without necessarily stopping or walking. As my friend and colleague Ed Hay points out so well in his book about JIT,* the U-line's magic is not in the shape but in the fact that bending the line in whatever way puts operators physically closer together without necessarily being so close that they hinder each other. Unstraightening a production line allows workers to swivel, pivot, and turn to do more, either to help coworkers or possibly even to do two or more operations at different stages of the process.

A U-shaped layout increases flexibility in staffing, so that each quarter, month, day, or even each shift, if the cycle time is changed, the staffing level can be changed. Each task is available to multiple operators working in areas, not just at one station along a continuous conveyor. Operators working inside a U-line or similarly laid-out line aren't constrained if they want or need to help out with the next operation up the line or with the previous operation.

Another good layout includes two lines back to back running in opposite directions. Again, work can be assigned so that some workers with light loads can do operations on both lines.

If employees are to have maximum flexibility to work in the entire process area, or at least a big part of it, it will be necessary to remove the large hunks of work-in-process (WIP) inventory that often get stacked between operations. The reduction and eventual removal of WIP is made easier by the very act of bending and curving lines and putting operators

***The Just-In-Time Breakthrough** (New York: John Wiley & Sons, Inc., 1988).

6-1. and 6-2. A straight-line production line layout inhibits flexibility, impedes communication among workers, and makes it nearly impossible to know when parts are needed and by whom. The line illustrated in these pictures is so long that it takes two photographs just to show its length!

Figure 6-7. The U-shaped production layout.

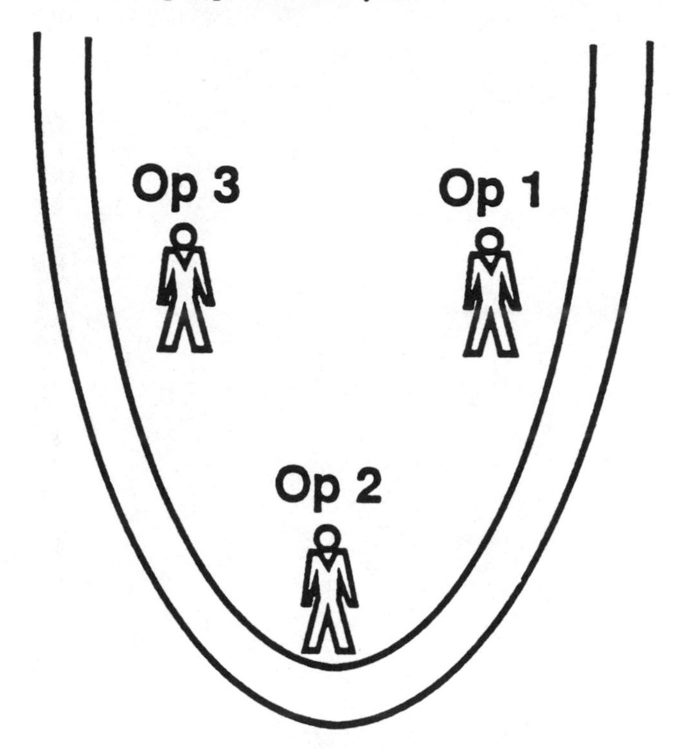

in motion to accomplish multiple tasks. Parts should be easily within an arm's reach. In a JIT layout, it is easy to see when someone needs help. If a person is absent, any of the other people on the line can fill in. If customer demand increases, requiring greater production and thus more people, the line is easily modified.

The line layout at Valleylab—improved by the employees—has incorporated the "U" concept as shown in Photo 6-3. Workers are paced not by the conveyor in this example but by the number of pencils on the tray. Backtracking should be avoided. Parts should always be flowing in one direction to avoid confusion.

This may seem like an assembly example, but it really doesn't matter whether these work stations are assembly, fabrication, or administrative. The principle still applies.

Linking Different Processes to Ensure a Continuous Flow of Product

Many traditional manufacturing environments are organized by process specialty rather than by product line. Assume that in one company circles are required before building triangles and triangles become squares

6-3. Workers at Valleylab were the force behind instituting the U-shaped production line shown here. Pacing is now driven by need for parts, easily seen at a glance, and any hitches in the line are readily detectable.

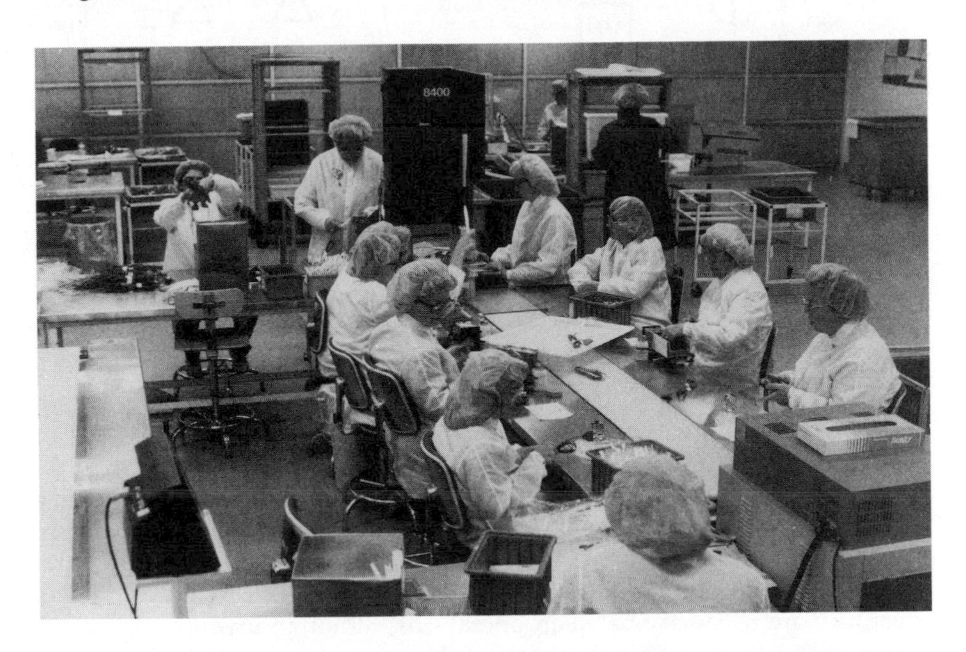

in order to build a widget. In Figure 6-8, representing a traditional setting, all circles are made in the circle department with a circle supervisor; the same is true for triangles and squares. Since each department is separate, work-in-process (WIP) inventory must be built between each one. Batches and piles of inventory are inevitable in this environment. Separate work orders are issued for each department with a specific schedule for start and completion. For instance, if each product takes one day of standard time to build, then a work order would be issued for circles to begin on Monday and finish on Tuesday. Another work order could then be scheduled for triangles to begin on Tuesday, assuming the circles will be done. Typical shop-floor scheduling works this way.

Since building product does not always begin and end according to standard, extra inventory between each operation is a given to make sure every operation has something to work on. Tracking of individual orders is necessary to understand what is being built and where.

These circles, triangles, and squares can represent machines, processes, or assembly work stations. Possibly milling, grinding, and drilling could have been the three departments. Processes such as mixing, forming, and coating in pharmaceuticals or prep, insert, wave solder, touch up, and test in electronic printed circuit boards follow exactly the same concept. In

Figure 6-8. Manufacturing environments organized by process specialty.

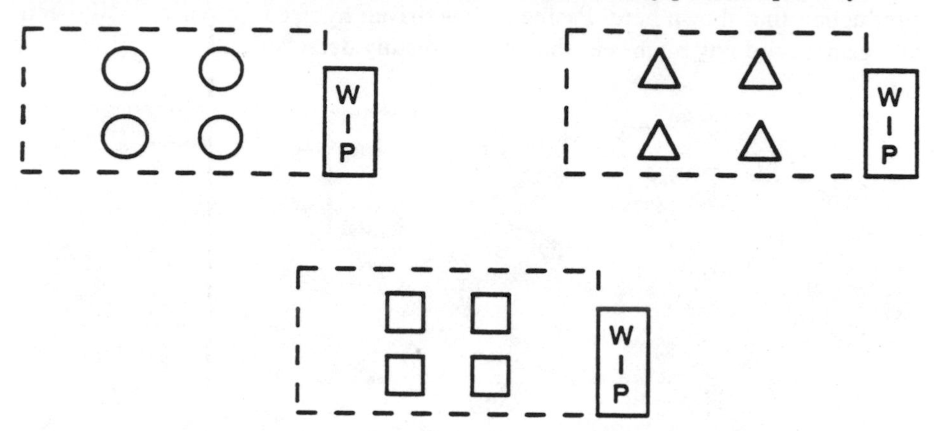

fact, what about the steps involved in processing a customer order or issuing a check in accounts payable?

In Figure 6-9, a circle, a triangle, and a square have been linked into a single department, with one supervisor, to make widgets start to finish. The layout now is conducive to communication among the three operations. One work order to build a widget can now be issued. No piles of inventory are necessary between operations, although inventory may still be required before or after this area until all processes are linked, supplier to customer.

The Ten Phases of Creating "Product Centers" (Work Cells) to Link Different Processes

When many different products are built on many different machines or by many different types of people and in different processes, it is difficult to know where to begin to create a flow of product through multiple processes. I use the term *product center* to depict a flow of product, start to finish, within the same area. Ideally, one team completely builds a particular product, creating a lot of ownership for the results of the final product. The term *center* was coined by a client and seems very descriptive of what is really happening. The way to best group products into centers may not be by product groups or families. A step-by-step process for analyzing products and processes to identify product centers and determine capacities could follow these ten phases:

1. Develop a matrix of products with processes/equipment/setup/machine man-hours (pieces/hr) for each. You may want to do a sort of your routings to get common processes/equipment.

Figure 6-9. Manufacturing environment organized for flow.

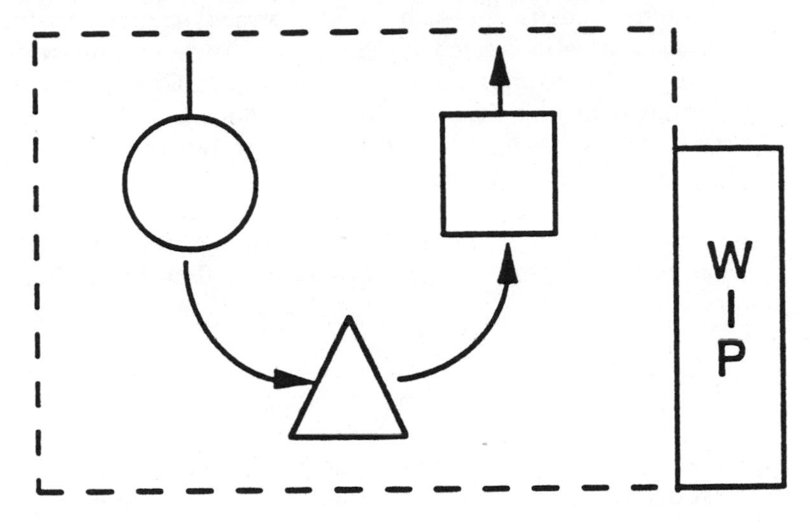

2. Find the products that fall together with like processes (not exact—just similar).
3. Determine the percent of product that has commonality.
4. Pay special attention to the 20 percent of products that make up 80 percent of the volume/dollars. For instance:

Product	*Op 1*	*Op 2*	*Op 3*	*Plating*	*Assy 1*	*Assy 2*	*Assy 3*	*Pack*
XYZ	X	X	X	X	X	X	X	X
A	X		X	X	X		X	X
B		X	X	X			X	X
C	X	X		X	X	X		X

Products XYZ, A, B, and C have similar processes, although not exact. They may fit into a cell, however.

5. Create a "misc" family for all those parts that are not like the others. This may include small runs or specials.
6. Determine the resources required for each family of parts according to forecast/history, hours, and setup time.
7. Compare resources (people/machines) to these requirements.
8. Develop a plan to begin grouping resources according to product centers.

9. As the centers are created according to families, determine a maximum capacity for each (fully manned) center. As Master Schedules are now created by center, you can determine feasibility of the schedule compared with actual output. Once a schedule is determined by the demand, the shop floor can decide how to order setups/parts for best utilization of resources. It will also be able to tell you if the schedule is reachable.
10. Examine simple, visible ways to determine when to build. (Chapter 7 describes some options for you.) The schedule will tell you what to build. This will create a smooth flow throughout the process.

Product Centers at Valleylab

At Valleylab, the electrosurgery generators used to be built in several stages. Printed circuit boards were built by one division, while the generators were assembled by another. The printed circuit board area was organized by process: all prep together, all insertion of components together, and so on. Each process area had a leader who had technical expertise in that particular process. Completed printed circuit boards were sent to the stockroom.

The assembly area was made up of individual work stations dedicated to certain subassemblies or final assembly operations, regardless of what generator was being built.

The test area had a different manager and was also made up of individual test stations. It took fourteen test technicians over two shifts to test and repair all the units that were assembled during one shift each day.

Since the production process was reorganized, all functions report to a single manager in one division. This group has recently decided to organize around five different product lines, creating teams for each product line, including every operation from the printed circuit board through final assembly and test. Now each team will own its product line process, start to finish.

By the way, during intermediate steps to get to this plan, all stocking of parts has been eliminated, WIP has decreased drastically, and about one half the number of techs are needed, on one shift only, to test and repair units. Several techs have received promotions and new opportunities thanks to these improvements.

The terms *group technology* or *machine cell* are older terms, used before JIT. The idea behind these terms is similar to centers up to a point—both encourage the grouping of unlike processes to build a product.

There are additional requirements on this concept for a JIT center, or cell. First, to be a true JIT cell, product flows one at a time from process to process. Secondly, a true JIT cell has the flexibility to operate at different output rates and different numbers of people to accommodate a changing customer demand.

Thirdly, in order to provide the opportunity for 100 percent involvement, the JIT center should be a *team* of people dedicated to producing product—start to finish—by flowing the product through multiple processes.

You may be thinking that this concept will require additional investment in equipment, people, or other resources. You may be right. However, the savings in the reduction of inventory and the increase in quality can more than compensate for the additional expenditures, if necessary.

Step 4. Develop a Plan for a New Layout That Promotes Flow

Brainstorm new ways to design the product flow; include the team that works in the area as well as engineers when needed. Let the team members develop ideas and then critique those ideas according to the six criteria mentioned earlier in this chapter, plus other criteria that may be important to them, in accordance with any specific boundaries given to them, such as facility or budget restrictions.

A paper model provides an effective tool that works great for exploring new ideas for layout. First, create a piece of paper that represents the physical room or area according to some simple scale (example: 1 in. = 1 ft.). Then cut out colored paper to scale for everything that requires space, including desks, chairs, equipment, tables, and any other necessary supplies.

Next, arrange this paper to represent the room with everything as it is now and place it on a large table that is accessible by the team and that will not be disturbed. The team may want to number the process steps within the layout according to the way the process is done. In other words, if step 1 in the process occurs at machine X, then put a 1 on the piece of paper that represents machine X.

Now the team can begin moving the layout to provide a simple flow, considering the six criteria mentioned earlier.

This kind of tool may seem too elementary, but it works. Everyone can get involved, and ideas can be tried as employees think of them. Again, visible and simple tools win out over complex ones.

After the team has developed its best solution, it is usually necessary to present this solution to management, but the team should be empowered to implement the changes.

Step 5. Implement the New Layout for Immediate Results

It is very important that the team be involved in every step of implementation. Even if some steps are being done by others, team members should know exactly what to expect and when. Later on, minor changes in the product flow—something as simple as moving a bench—may not require an approval cycle and implementation plan. Teams should not work around documented procedures but change the procedure first.

Let's talk again about ownership. Layout and product flow design have always belonged to management and engineering. So why does this have to change?

Consider for a moment how you might feel if one morning you arrived at work to find that your entire area had been rearranged. Even your own tools and personal things had been moved. How would you like it? It is virtually certain that you would not agree with the changes, nor would you like it! However, if you have been part of the process and your ideas have been included, your attitude is likely to be totally different. Involvement makes change easier. Having no surprises makes change easier. Commitment to the change has far more impact on the success of the change than the actual design does. Therefore, the watchword is to take a little longer and initiate change guaranteed to succeed.

Step 6. Continue to Evaluate the Flow for Improvement

Continuous improvement applies to the product flow. Once a new layout and product flow are implemented, problems will surface that were hidden before. A constant evaluation of the product flow that cycles through Steps 1 through 4 will ensure success.

How to Avoid Automating Wasteful Processes

Books are constantly written about automation, so I do not discuss the subject in detail here. Besides, in many ways, automation is like scheduling in that if you have it in place, you shouldn't take it down right away because you are implementing JIT; if, however, it is not in place, you need to think through carefully how much automation you really need and where you need it. Automation does no good if you merely automate wasteful processes. First, get rid of the waste; then consider automation.

The Impact of Automation on Better Product Flow

Automation is a controversial subject. In today's world, many consider automation the way of the future. People love the challenge of designing and implementing automation. And, in fact, automation is the answer for success in some cases. However, it is important also to consider the ways in which automation might limit continuous improvement of your product flow.

Automation can include not only the automation of processes in nonmanufacturing, but also the automation of material handling and information (usually with computer systems).

The Pros of Automation:

- Better quality, especially for boring jobs
- More consistent than people
- Faster, stronger, and safer for some jobs

The Cons of Automation:

- Usually more complex
- Limits flexibility
- Institutionalizes processes, procedures, and data
- People see problems and possible solutions machines don't see
- The more machines, the more possibility of down time and the greater the unpredictability of when that will occur
- High maintenance costs may not be apparent early on

The Right Timing for Automation

Assuming that automation is appropriate, the next question is how to time its introduction.

In order to automate production, material handling, or information effectively, these steps must be followed:

1. Document the process.
2. Perform a value-added analysis, identifying those activities that do not add value.
3. Eliminate as many nonvalue-added activities as possible.
4. Evaluate the "new and improved" situation, and perform a cost/benefit study for automation within that situation.
5. If the benefit is greater than the cost, or if there is a significant benefit in quality, then automate.

This last point is important. It is difficult today to justify automation strictly in terms of labor savings, since labor is such a low percentage of total product cost. Sometimes a careful analysis will reveal that not only is the timing not right for automation but that automation is not wholly justified. JIT is a process of change and continuous improvement. Automation at the wrong time can prevent progress with JIT by institutionalizing wasteful processes and limiting flexibility.

Creating an even, balanced, and self-adjusting flow of product through the manufacturing process is an important first step inside the factory. The mindset of continuous improvement will mean continuous ideas for improvements to the product flow. The steps of thinking flow, documentation, critique, and implementation can be repeated over and over in all processes—manufacturing or administrative.

Chapter 7

Four Steps for Implementing Demand Pull Successfully

Once the process flow has been improved, the method of moving material through the process can be examined and improved. Once again, the analogy of city traffic can be used; once the city's road pattern has been laid out, it is time to regulate traffic movement by putting up street lights, signs, and other devices.

The best principle for regulating material traffic in a manufacturing environment is demand pull, otherwise known as a pull system or by the Japanese term "Kanban." Kanban is used by Toyota as the signal that tells a manufacturing process that more product is required. Toyota uses cards to signal this need for more material; in fact, one translation of "Kanban" is "card." However, Kanban has also been used to refer to the entire principle of demand pull. Therefore, the word "Kanban" has long been a source of confusion in many American companies. I refer to the signal sent from one point of manufacturing to another for more material as the demand signal.

In a traditional manufacturing environment, material "pushes" its way through the system; each operation is issued a schedule based on capacity, a batch size, and a standard time for completing the batch. Each operator works independently to satisfy the requirements of the schedule. As long as everything is within the standard time, parts flow smoothly through the process.

But if problems occur at one operation, inventory quickly builds up because previous operations continue to build as long as parts are available. Each operation is unaware of what is happening at other operations except through schedule adjustments made to react to an inability to meet the original schedule. Because of time lags in communication, by the time everyone is working at the new schedule, the manufacturing scene can look like Figure 7-1.

Photo 7-1 shows such a situation at Valleylab before JIT; parts have piled up because of an uneven flow of product.

Once the problem has been corrected, the operator in the affected area will have to work overtime to catch up on the huge pile of material completed by previous operations. If there is a quality problem in that batch, the whole overbuilt pile may need to be scrapped or reworked.

There are four steps to take in order to convert from this "push" method of moving material through a manufacturing operation to a "pull" method. Under the pull method, the final operation must actively signal a previous operation that it needs more product in order for the previous operation to build more. Demand pull provides a simple and visible way to allocate resources to provide what customers want, when they want it.

In this chapter, I compare "push and pull" and then discuss the four implementation steps for demand. Finally, I compare traditional shop-floor control with demand pull.

A Comparison of Push and Pull

Pull scheduling provides immediate feedback if the flow or product is interrupted. You have the opportunity with pull to apply resources where

Figure 7-1. Impact of operation stoppage on other operations.

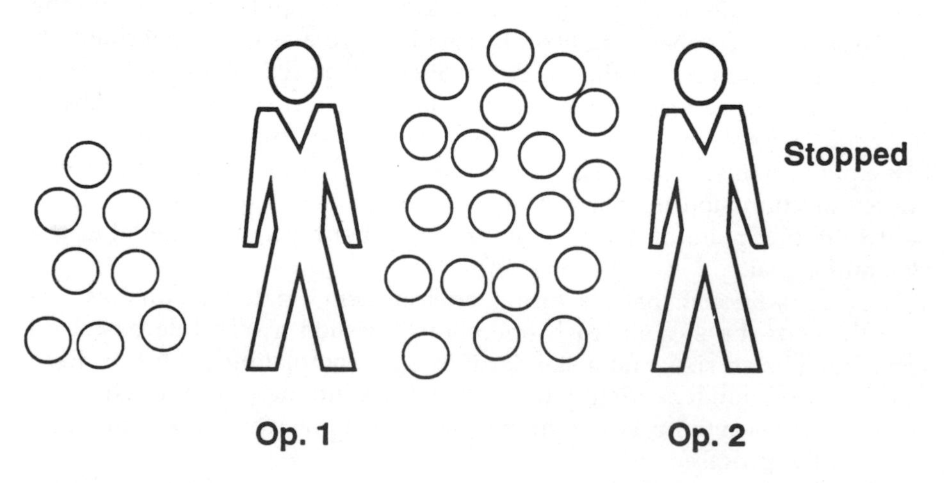

7-1. Many generators have piled up at this operation, indicating an imbalance in the flow.

they are needed most to meet the customer requirement. A self-managing system can be developed with pull scheduling that is synchronized with the customer (see Figure 7-2).

The benefits of demand pull include:

1. Team ownership of the flow
2. Visibility
3. Simplicity
4. Creation of an urgency to solve problems
5. Proper allocation of resources to work on the right things to satisfy customer need

The four implementation steps can be utilized repeatedly as the process flow improves. In order to pilot demand pull, note the following points:

1. Clearly define the beginning and the end of the flow. You can think of it as a race with a start line and a finish line.
2. Assume that material will be piled just as always before and after the defined flow.

Figure 7-2. A comparison of push and pull.

Issues to Consider	Push Scheduling	Pull Scheduling
Responsibility to monitor	Scheduler/system	People in manufacturing
Flow control	If all standards are met	Always
Signal to build	Schedule/system	Visible - from customer
Inventory	No limit - higher	Limits - lower
Problems	Can be hidden	Exposed - creates urgency
Communication between operations	By chance - operations work on thier orders	By necessity
React time: Changes/ problems	Through the system - rescheduling required	Immediate - on-line and visible
Visible indication of problem	Inventory	Not working
Shop floor control	System, transactions, paperwork	Automatic - visible and simple

3. Learn to manage demand pull within the flow and then expand the scope of the pilot.

Now let's explore each of the four implementation steps carefully.

Step 1: Stabilize Queue Sizes to Limit Parts Pileup

Let me first clarify what I mean by several important terms in this discussion:

Queue refers to the amount of material between operations. Of course, the goal of JIT is to eliminate queues by balancing the flow of material throughout the process. But as an intermediate step—so no customers will be hurt while a company implements JIT—the queue sizes can be stabilized and reduced.

Move quantity refers to the amount of material being built and moved as a batch. Notice in Photo 7-2 that the build quantity is very large at this work station. If there is a quality problem, this entire group of parts will have to be reworked.

Order/lot quantity refers to the group of parts that is authorized to be built. Usually these work orders/job orders/lots are created by production control and provide a way to track a group of parts through the manufacturing process.

7-2. Batch quantity is large at this work station. If any quality problems in this batch are detected at a later stage of production, much time, effort, and money will go into correcting the defect.

The order, or lot, quantity may or may not be the same size as the move quantity. A lot may be 300 pieces, but it may be built and moved in batches of ten until the entire 300 are built and moved and a new lot is started. The primary goal of JIT is to drive the move quantity closer and closer to one-at-a-time—build one part, move one part to the next operation, etc.; a secondary goal is to bring lot sizes down. Lot sizes can be reduced by setting up jobs and changing over machinery faster than before; however, the move quantity to one is primarily a matter of people developing a mindset that says "one-at-a-time flow."

The first step in implementing demand pull is to look for places in the process where one-at-a-time can be built and moved, thus eliminating queues. It is important in changing the mindset to show that one-at-a-time can work; to do this, problem-free areas must be found to try the technique in pilot projects. At the same time that material in one area, as small as it might be, is being moved one at a time, the other queues throughout the process should be stabilized.

> The key is not to eliminate all queues as a first step. Rather, it is to establish planned, carefully thought-out queues with specific demand limits.

A good rule of thumb is to stabilize the queue at a quantity that is comfortable for today's process. The goal at the beginning is not to reduce the queue but to establish a limit for the amount of material that can be in each queue.

Establishing a Queue-Size Limit That Can Stabilize the Process

The first requirement for establishing the correct limit for queue size is to get the team involved. The people who build product in the particular area understand best how the amount of material fluctuates and where the line imbalances typically occur.

Second, determine the minimum, maximum, and typical queue sizes that usually occur in the process. Also determine the batch size necessary for the subsequent operation.

Third, determine the lead time necessary to replenish the queue in order to have product flow continuously. (Many people mistakenly assume that the queue size should be tied to the volume of product demand; the queue increases as demand increases. This is incorrect. An increased demand means that the speed at which the queue is filled and then emptied may increase but not that the queue needs to be larger.)

When all these items have been considered, the team needs to determine a queue size that it is comfortable with and adhere to it; regardless of how long an operation ahead is down, if the queue is full, prior operations must also stop (Figure 7-3). That's a big cultural change from a build-as-long-as-you-have-parts mentality.

Figure 7-3. Filling the queue.

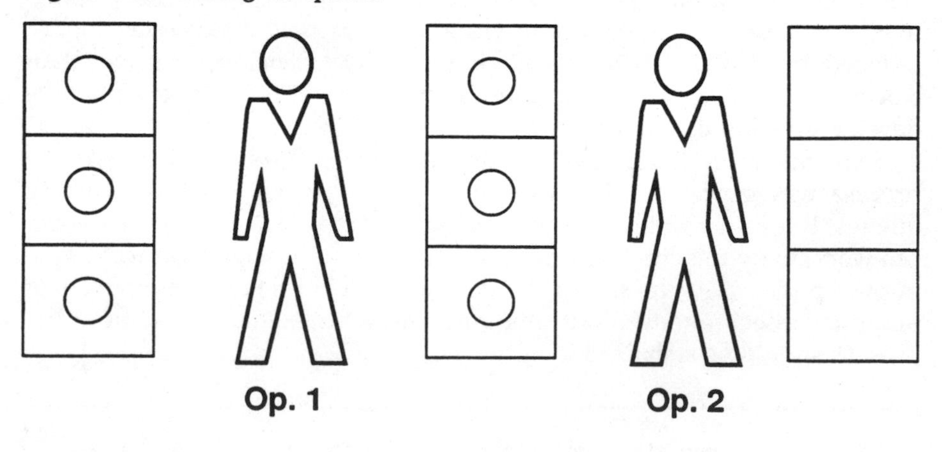

Step 2: Stop for Problems to Avoid Building Bad Quality or the Wrong Thing

Setting a limit on queue size sends the message that it's okay to stop. Problems in a process may indicate a quality problem or a line imbalance. In either case, the "customer"—the operation ahead—is not ready for more product. I have discussed in Chapter 2 the cultural change required to stop the process and the implications of appropriate stopping for quality.

The costs of continuing to build despite problems in the process are quality and flow; the benefits are keeping people and machines busy. The correct conclusion is that the benefits are not worth the costs. However, in the past, this was rarely the case; rewards and incentives were almost always based on personal production quantity.

Step 3: Implement Visible, Simple Demand Signals

Once queue sizes have been stabilized and people have gotten used to stopping for problems, it is time to shift the focus of moving material from pushing material forward to pulling material through the process as it is needed. To do this, team members need to agree on appropriate signals for the "customer" operation to use in order to tell the "supplier" operation to produce more and put it in the queue.

The signals used to indicate that product is needed should be visible and simple; possibilities include squares marked off on a table, specific bins for specific items, colored circles on a board, or indicator cards. The signal for units going into test at Valleylab is a marked shelf (Photo 7-3).

When we were first learning how to implement demand pull at Hewlett-Packard, we didn't keep it simple. Before we knew it, we had engineers writing software to indicate demand at each work station, which required each work station to have a terminal and some pretty complex software (as well as a number of nonvalue-added steps).

A good test of the simplicity and visibility of demand signals is to bring someone into the area who is not familiar with the process and ask that person to explain the signals after merely watching the process for a few minutes.

Step 4: Implement Demand Pull to Allocate Resources Where They Are Needed Most

> Demand pull always tells when to build, and sometimes what to build, too.

7-3. Some sort of easily seen signal is helpful in establishing demand pull in order to let workers know when more product is needed. Use of a marked shelf, as here at Valleylab, is one possibility.

There are three simple rules for working in a demand-pull environment:

1. Don't start any work unless the demand signal indicates a need for more material.
2. If the demand signal indicates a need, work to fill that need.
3. Never exceed the queue-size limit.

Let's explore how this would work by adding a material staging area to our example that uses operations 1 and 2. The demand signal chosen by the team running this operation for communicating "more material is needed" is an empty square. The queue limit is three, indicated by three squares.

In Figure 7-4 the material has been delivered to the staging location. Operator 1 should begin to build, because the demand signal indicates a need for material. As he takes material from the staging location, more material is delivered to fill the demand. Operator 1 stops building when he reaches the queue limit of three (see Figure 7-5).

Figure 7-4. Filling the queue: beginning to build.

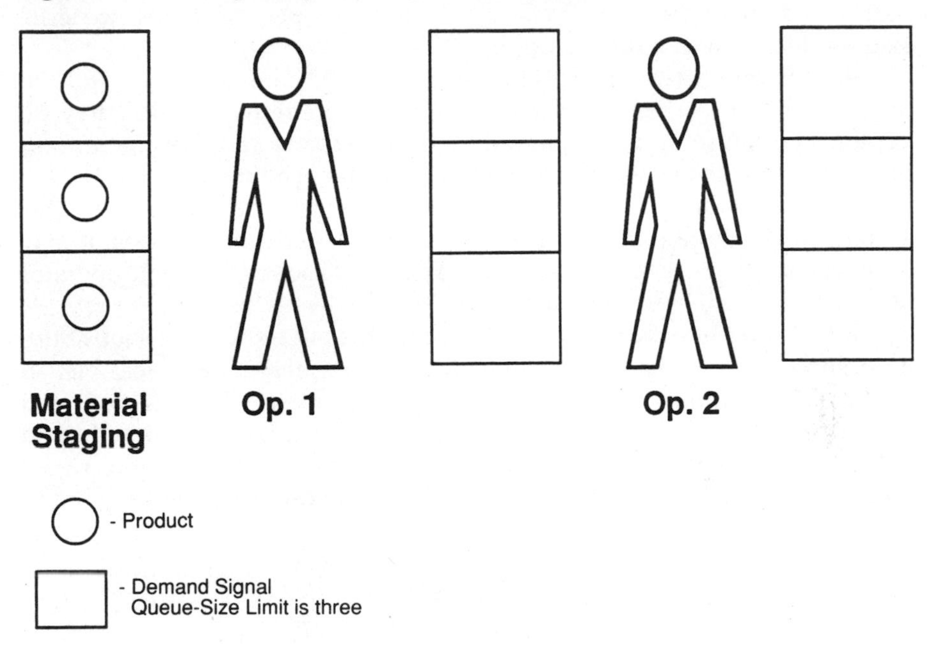

Figure 7-5. Filling the queue: demand-pull production.

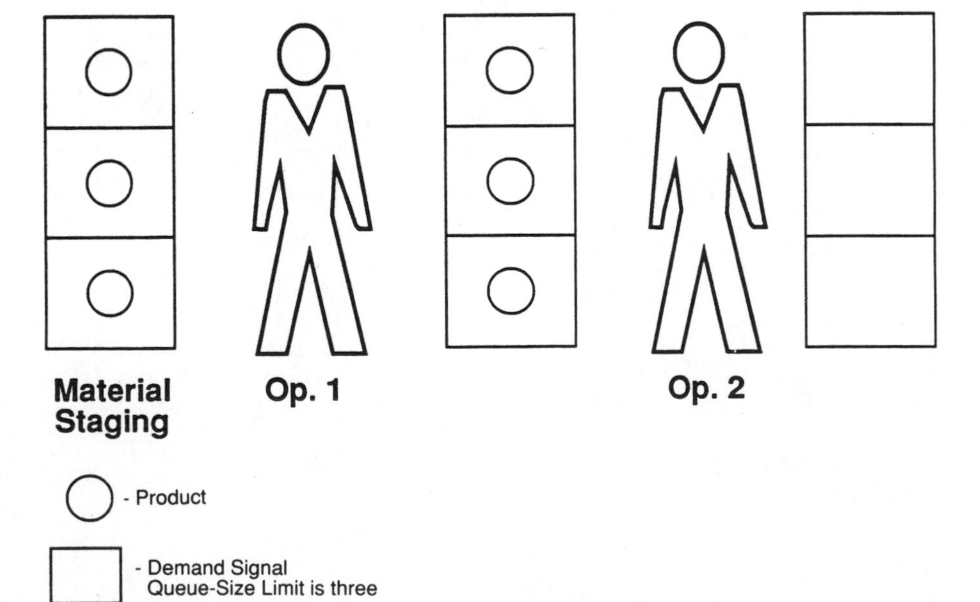

Operator 2 should now be building, because his demand signal indicates a need for more material. However, operator 1 and material staging do not need more material.

At this point, team members can build in batches of three to fill the queue or build in batches of one to keep the queue full. Which they do depends on the setup time required for each part and whether the process has been perfected to the point that building one at a time is feasible.

If there is only one product variation on this particular line, this system works quite well. But if there are many product variations, it may be necessary to label the squares for specific products so each operator also knows which ones to build (Figure 7-6).

If the number of variations is too great or, as in the case of a job shop, each successive product is different, this scheme may not be feasible. In either of these situations, it may be necessary to use a schedule generated from customer orders or planned demand to determine *what* to build and to use demand pull to determine *when* to build. The demand signal indicates a need to build something, and the schedule determines what to

Figure 7-6. Demand signals labeled for different products.

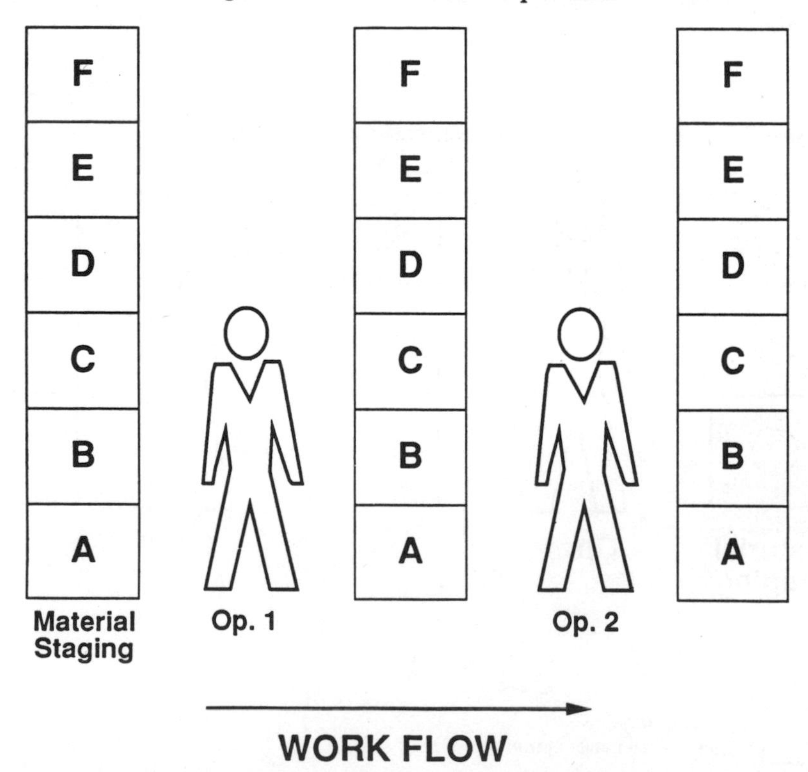

build. This demand signal could represent a certain amount of work in quantities of parts, labor required, or other categories that make sense.

For instance, different products in a product family may take different lengths of time to build. In this case, the number of parts would not necessarily lead to capacity usage (Figure 7-7).

How Demand Pull Replaces Shop-Floor Control

Shop-floor control is a term used for a system of tracking material in and out of operations in order to provide visibility of material, and sometimes

Figure 7-7. When and what to build.

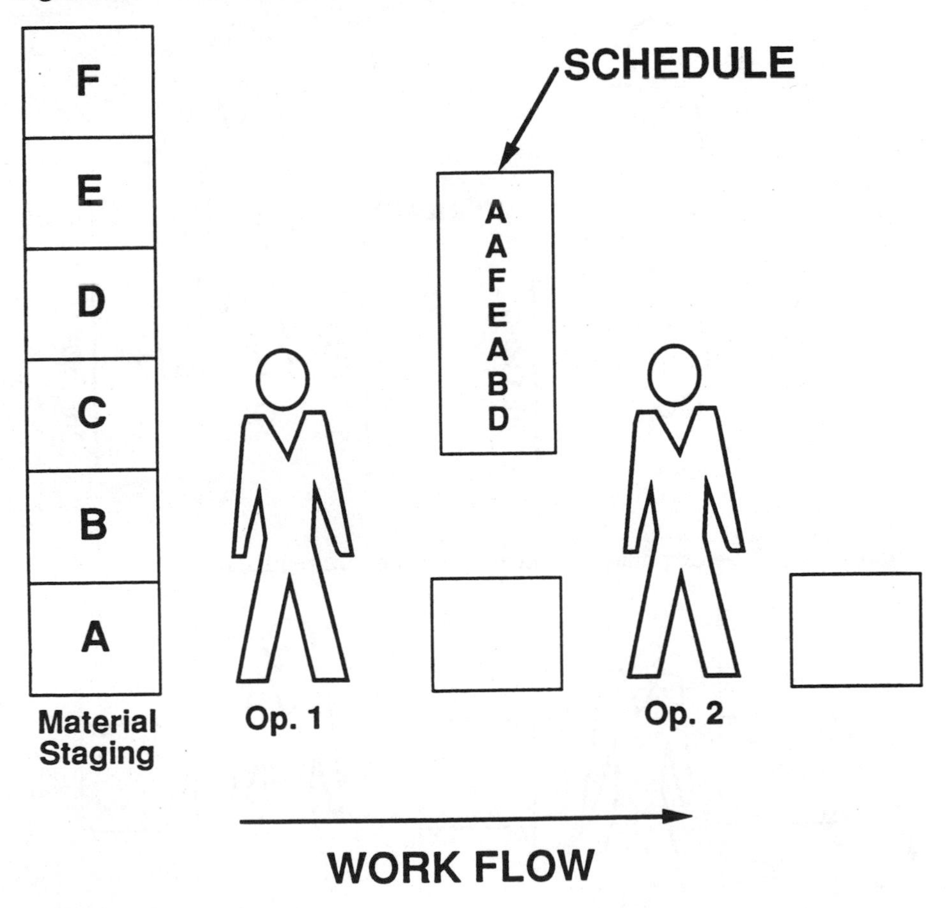

scrap and labor. Usually this requires a computer system. In Figure 7-8 shop-floor control is necessary to track material into and out of each operation.

As operations 1 and 2 are linked through demand pull, the cycle time decreases considerably. No longer are there unlimited queues to get through, and setting priorities is automatic. The next step may be to track material only into this flow line and out of that line, as shown in Figure 7-9.

The next step is to link these operations to every other operation required to build the part start to finish. Cycle times become consistent because of the creation of a Drumbeat, flow, and demand pull. No longer

Figure 7-8. Shop-floor control tracking.

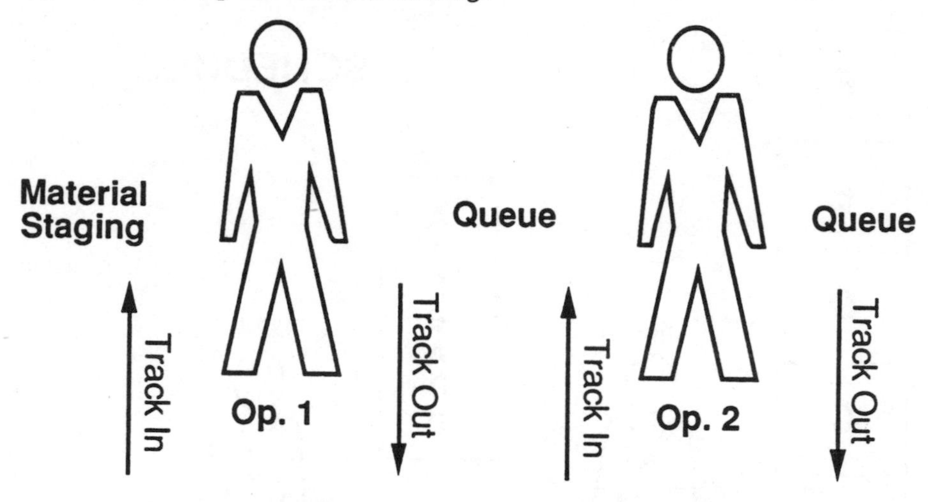

Figure 7-9. Tracking material when cycle times are reduced.

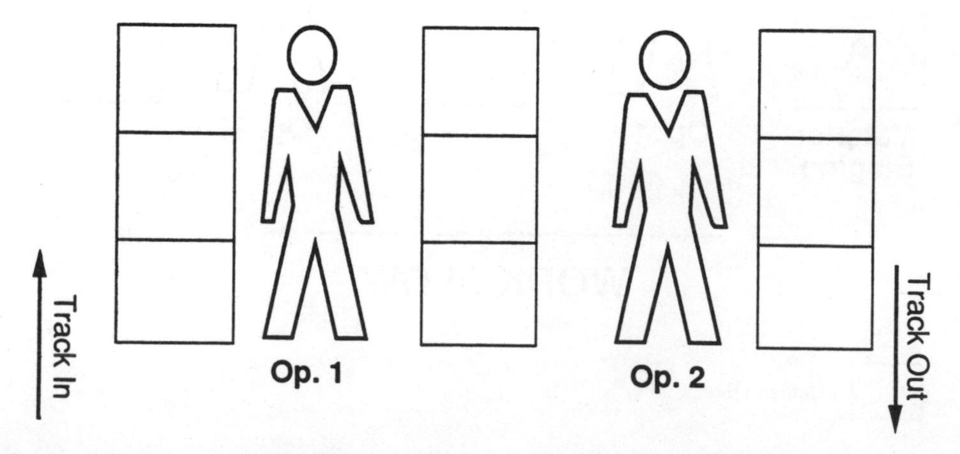

is there a need to track anything; once the product is started, it is easy to figure out when it will be completed.

If you don't already have a shop-floor control system in place, I would advise you not to implement one. As you progress with JIT, the chances of needing one will be small, and if one is needed, the requirements will be very different from the requirements you might have today.

The technique of demand pull works in any environment. However, the implementation steps and challenges may differ widely, depending on your products and processes. Remember, there is no right place to start a demand-pull regime; begin where it makes sense and expand with success.

Chapter 8

How Inventory Reduction Results From Success With JIT and Total Quality

As you may recall, Just-In-Time is *not* an inventory reduction effort—instead, inventory reduction is a result of success with several JIT techniques, including flow and demand pull. JIT and Total Quality are inseparable; JIT exposes problems, and a quality effort solves those problems. JIT and TQ are linked through the techniques for reducing inventories at a pace at which problems can be solved.

In this chapter, I first discuss how inventory becomes a liability instead of an asset. I move on to the advantages of having less inventory and finally the progression of inventory reduction from work-in-process (WIP) to raw materials to finished goods.

How Managing Inventory as a Liability Can Drive Inventories to the Right Level

Inventory has always been considered an asset in any organization—in fact, the balance sheet shows it as such. Profits on a profit and loss statement can actually be inflated by carrying more inventory. But in reality, too much inventory is a liability. The advantages of carrying only the right amount of inventory are significant. By thinking of inventory as a liability, you begin to look for opportunities to reduce inventory.

Let's first explore some traditional thinking about inventory and how JIT changes attitudes about inventory.

- *"Just in case."* Traditionally, people have believed that inventory provides security "just in case" something goes wrong. If the materials department is reprimanded for running out of parts once, it orders twice as much just to be sure that doesn't happen again. JIT simplifies planning and scheduling to promote accuracy, allowing for only the right parts to be in-house as needed by manufacturing. The materials department is part of the manufacturing team.

- *Job security*. People have become accustomed, over the years, to always having piles of inventory to work on. It is frightening for them to have only what is needed. Many people feel insecure without the pile. They must get reassurance of more work to come under JIT by understanding the advantages of less inventory and better flow.
- *Automation*. Material storage and retrieval systems have been in vogue for some time now. The real question to ask is: Can we eliminate some waste and therefore the need to automate?
- *Quality*. Traditionally, it has been felt that high inventories are needed just in case there is a quality problem—if a part is bad, another one is available right away. JIT forces you to think of 100 percent quality the first time.
- *Keeping busy*. Extra inventory in manufacturing areas in which lines are not balanced well or in which process or equipment problems are frequent has always been essential to keep everyone and every machine busy. With JIT, it is okay to stop for problems and be busy only when it is possible to build quality products that fill a demand from the customer.

Thinking of inventory as a liability requires training and a lot of reinforcement. In order to change that kind of thinking, everyone needs to understand why just-in-case inventory is wrong. Those who have accounted for inventory as an asset in the past are required to change their way of thinking, and those who previously were not made aware of inventory must be made aware of it and of the consequences of having too much. Having the right amount of inventory is everybody's job.

Three Advantages of Having the Right Amount of Inventory

The idea of "zero inventories" is really a misnomer; the goal is to have the *right amount of inventory* on hand to *build the right product* for the customer at the *right time* with *100 percent quality*. Having just the right amount of inventory has three advantages.

1. Reduced Costs and Increased Cash

Inventory ties up cash and creates expenses that add up quickly. When developing a cost and benefit model, don't use the dollars tied up in inventory as savings—top management will reverse that one every time. Simply state it as it is—a freeing up of cash.

There are also some actual expenses incurred when inventory is carried that can be justified as real savings, however. An accepted amount of expense in most companies today for carrying inventory is 30 percent of the value of the inventory. You can use the worksheet in Figure 8-1 to calculate the cost of carrying inventory.

In an effort to cut handling costs, some companies have gone to capital-intensive and expensive automated inventory movement and storage equipment. The right question, however, is not, "How can we automate our inventory?" but "Why do we have so much inventory that we need to think about automating it?" This is another case of automating waste!

It is important to establish the inventory carrying cost percentage in your company early in the Just-In-Time implementation, since this is an easily measured savings as inventories are reduced.

2. Better Quality

Let's go back to the two operations in Chapter 7. In which scenario would quality problems be noticed more quickly?

As we discussed in Chapter 4, producing fewer parts at a time makes it easier to focus on quality and get resolution to any problems; as WIP inventory is reduced as a result of a better flow, quality always improves. Focus, feedback, and resolution occur when the number of parts is smaller.

3. Reduced Cycle Time

Cycle time refers to the *elapsed time* from the start of a manufacturing process to the completion of that process. This process can be defined by you; it might be several assembly operations or an entire product line flow from beginning to end. WIP inventory reduction will occur as cycle times are reduced. There is less waiting in queues, fewer large batches, and better flow.

The way in which these three factors—inventory, quality, and cycle time—correlate is something everyone trying to create a JIT environment needs to understand.

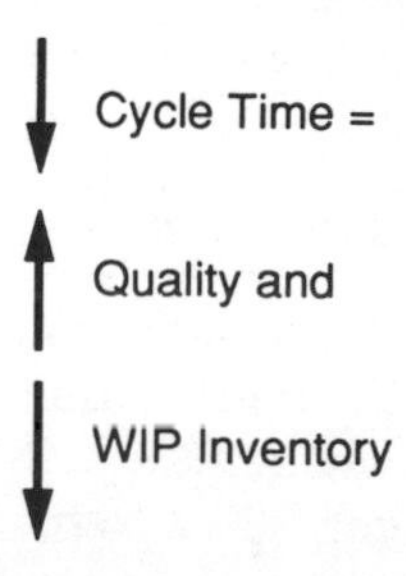

Higher Inventory Leads to Greater Problems

That challenging drive through the city of Boston comes to mind. The expressway has three to four lanes in each direction that are almost always filled with vehicles. Let's consider the traffic from one exit to the next.

If vehicles continue to squeeze in from one exit to the next, will the vehicles get through more quickly? Of course, the answer is: No, they'll probably go even more slowly. So the more vehicles, the longer the cycle time, or elapsed time from one exit to the next.

What about the chances of damaging my car as I squeeze in closer to allow for more vehicles? The chance for damage or confusion is greater. Therefore, as inventory increases, problems with quality are more likely.

As cycle times are reduced, responsiveness to the customer increases. It is always easier to predict demand over a short period of time than over a long period; uncertainty increases over time. As a company reduces cycle time for a product, it is able to be more a build-to-order operation and less a build-to-forecast operation that must hope the forecasts are right.

How to Reduce Inventory Systematically as Problems Are Solved

In reducing inventory, a cardinal rule laid out by my colleague Ed Hay is:

> NEVER LET THE CUSTOMER GET HURT WHILE YOU ARE LEARNING JUST-IN-TIME.

In order to reduce inventories without taking the chance of hurting the customer, each step needs to be thoroughly thought out.

Everyone who has read about JIT has heard the "ship and rocks" analogy. Let's use it to illustrate how problem-solving speed paces inventory reduction.

In Figure 8-2, the ship is floating merrily along, oblivious to the rocks under the water. These rocks—hidden problems such as supplier quality and delivery, design problems, system problems, and process problems—are hidden in an ocean of inventory.

If the water level is lowered, a rock will be exposed; now the ship's captain cares about this rock. The boat may be damaged if the rock is not removed. If the goal of JIT were inventory reduction, the water level could be lowered all at once. But the rocks would all be exposed, and the ship would probably get wrecked (see Figure 8-3).

(text continues on page 130)

Figure 8-1. Calculating cost of carrying inventory.

INVENTORY CARRYING COST WORKSHEET

Investment:

Cost of Money - Interest ______%

Transactions to Keep Track of Inventory:

	Quantity	Cost Per Transaction	
Receiving:	________	Time:	$________
Inventory:	________	Labor:	$________
Tracking WIP:	________	System:	$________
Costing:	________	Paperwork:	$________
Inspection:	________	Delay:	$________
Rejects:	________	Other:	$________
Finished Goods:	________		
Other:	________		
TOTAL # =	________	TOTAL	$________

#________ x $________ = Total Cost Today

TOTAL $____ / INVENTORY $____ = ______%

Receiving and Storing:

Labor (Salary & Benefits)	$________
Space (cost per sq. ft. × sq. ft.)	$________
Insurance	$________
Taxes	$________
Costing	$________
Counting	$________
Shrinkage	$________
Obsolescence	$________
Transportation (labor, vehicles, time, delay)	$________
Other	$________
TOTAL =	$________

TOTAL $____ / INVENTORY $____ = ______%

Quality:

Inspection $______

(Receiving and Manufacturing)

Labor

Delay

Rejects $______

Labor

Delay

Shipping

Rework $______

Due to Changes

(Engineering changes, Obsolescence, Problems)

Others $______

TOTAL = $______

TOTAL $____ / INVENTORY $____ = ______%

Other: $--------

TOTAL $____ / INVENTORY $____ = ______%

Investment ______%

\+ **Transactions** ______%

\+ **Receiving and Storing** ______%

\+ **Quality** ______%

\+ **Other** ______%

= **Total % of Inventory $s Expensed Due to Carrying Inventory** ______%

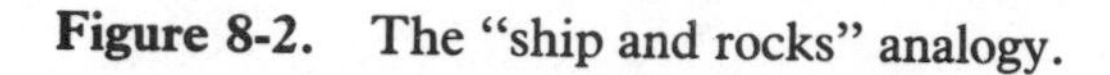

Figure 8-2. The "ship and rocks" analogy.

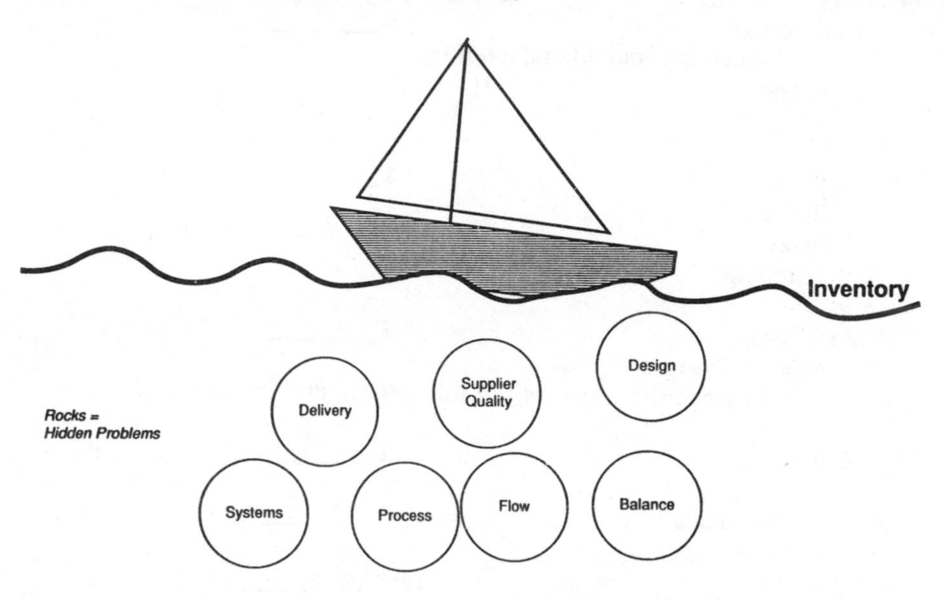

If inventory reduction is a goal, the organization will be driven to do the wrong things. Many managers use such ultimatums as, "I want the inventory to be at $100,000 by April." Materials people can accommodate this kind of a demand—one way or another—if it makes the difference in their success and their performance report. However, working toward a bottom-line goal does not necessarily get people to confront the issue of what properly should be held in inventory. Some necessary things may not be ordered because too much unnecessary stuff is on hand.

To put it simply, Inventory reduction is not a *goal*. Inventory reduction is a *result* of solving problems, and the speed at which inventory can be reduced depends on the speed at which problems with quality can be solved. The speed is neither constant nor predictable. But if everyone takes ownership for problem solving and has a continuous-improvement mind-set, the chances of ultimate success are very good.

How Reducing Inventory Can Reveal Other Manufacturing Problems

A debate comes up from time to time that I call the scuba diver debate. The critics of lowering the water level ask, "Why cause all the pain? Why not hire a scuba diver to solve the problems under water?" To that question I respond, "We already know where the rocks are, and we haven't solved

Figure 8-3. Dangers of reducing inventory too rapidly.

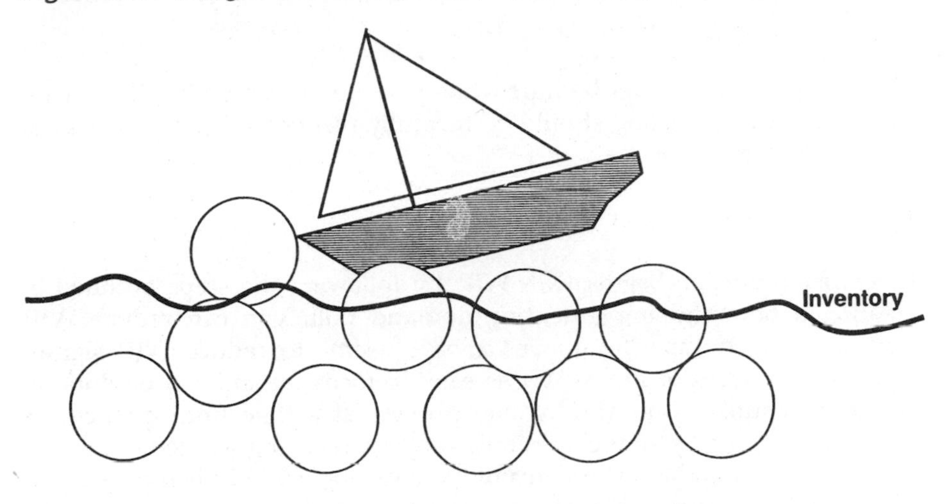

the problems with quality yet." The fact is that unless you disrupt your journey by deliberately causing pain, you have no incentive to solve the problems. They will just stay hidden.

This brings up an important point. An attitude needs to be instilled that supports the identification of problems with quality or flow or anything that inhibits customer service. In the past, workers have been sensitive about identifying problems because they are afraid that they will be blamed for the problem. In reality, though, these problems need to be identified and solved without recriminations. Problems must be seen as opportunities.

Eliminating Blame

On the final assembly line in one company, people were very hesitant to point out problems. A traditional, authoritative style of management had prevailed in the past, and sometimes people were questioned or blamed for problems when they identified them.

After some training, it was apparent that this style had to be changed.

In an effort to begin to turn this attitude around before trying to implement JIT, managers had a good and simple idea. They announced that there would be a new competition to see who could identify the most problems. Every time a problem was identified, the person got a "badge" on his or her smock sleeve and the manager came out and congratulated the "badge" recipient. This simple approach began to take away the fear of bringing up problems. Problems were now seen as opportunities.

How to Phase Inventory Reduction Without Hurting Customer Service

The customer must never be hurt while you are implementing JIT. Therefore, inventory reduction should be carefully planned to minimize risk of lowering customer service.

From WIP to Raw Materials to Finished Goods

Inventory reduction begins with WIP. By following the steps outlined in creating a flow and implementing demand pull, you can reduce WIP. Putting limits on specific queues always seems to reduce WIP significantly—50 percent or more. WIP is easy to focus on and can be done in bite-sized chunks—one area or one process at a time until queues are driven to zero, and WIP includes only those parts being worked on.

Figures 8-4 and 8-5 show inventory levels before WIP is reduced and after. As WIP is reduced, parts continue to come in from suppliers at the same pace. Inventory in raw materials therefore increases. This raw material inventory becomes visible through the materials requirements planning (MRP) and inventory systems and can thus be managed more effec-

Figure 8-4. Inventory levels before JIT.

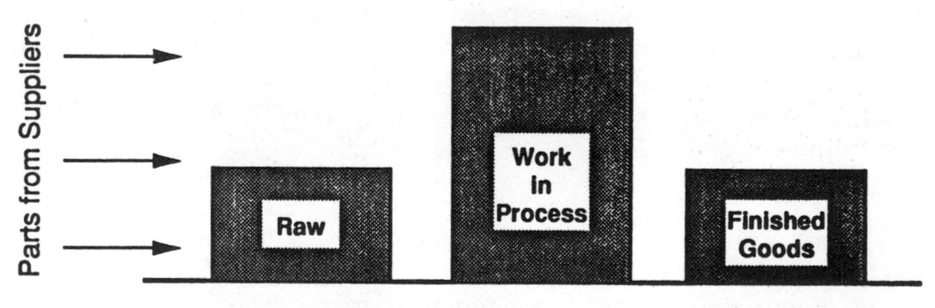

Figure 8-5. Inventory levels after JIT.

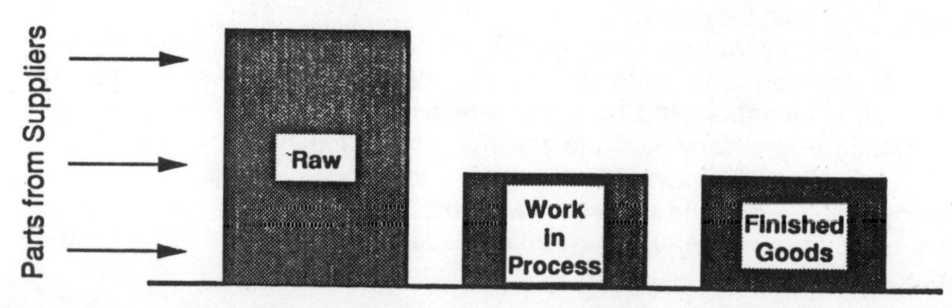

Figure 8-6. Reducing finished goods inventory with JIT.

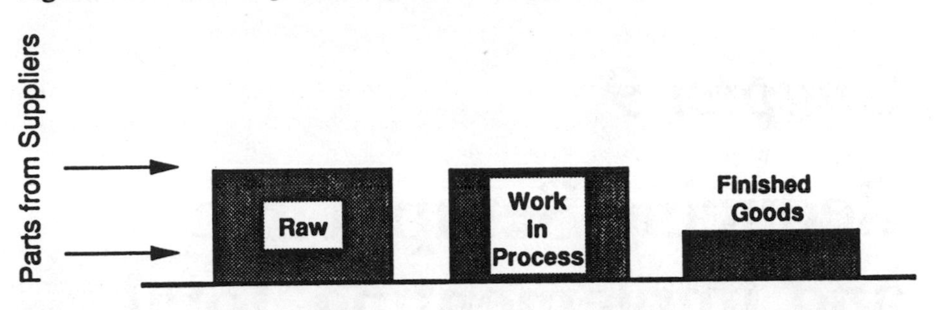

tively. Chances are greater that the right material can be made available at the right time. (I discuss this further in Chapter 12.)

Figure 8-6 shows that finished goods can be reduced while you maintain customer service.

Eventually, the focus will shift to finished goods, if that is a part of your business. If quality is increased and cycle time reduced, it is no longer necessary to carry a high level of finished-goods inventory. As you are able to build product as it is needed, and not before, within an acceptable cycle time, you can begin to dismantle shelving used to store unnecessary product in a build-to-store operation.

JIT forces a change in the mindset about inventory. Inventory reduction, done for the right reasons, can expose problems and create a sense of urgency to solve them. Just-In-Time and Total Quality are linked through techniques to reduce inventories at a pace that ensures problem solving and customer service.

Chapter 9

Reducing Setup Time and Implementing Total Productive Maintenance

Setup time reduction and Total Productive Maintenance (TPM) have been associated with environments with lots of heavy equipment that require lengthy changeovers and specialists to maintain the equipment. But if your company makes more than one variation of product in the same manufacturing area, you have an opportunity to reduce setup times significantly, and if you utilize any tools or equipment—large or small—TPM can help stabilize the process.

Most batch sizes in a traditional manufacturing environment are determined by the time (and cost) of the setup—not by what the customer requires. In fact, the tendency is to build a large batch all at once to take full advantage of the setup, even if a part of the batch is to be stored instead of shipped. I have discussed the cost of storing inventory in Chapter 8. Economic Order Quantity (EOQ) is the formula actually used to determine a cost-effective batch size; EOQ assumes setup time as a constant. But JIT drives the setup time in the EOQ formula to be a variable in order to build what the customer requires in a timely and cost-effective manner.

In this chapter, I outline the tried and proven steps to reduce your setup time by 75 percent at low initial cost and no cost after ground rules for setup reduction are established. By cycling through this process, you can achieve this 75 percent reduction over and over again.

The second part of this chapter focuses on implementing Total Productive Maintenance. TPM includes not only breakdown maintenance and preventive maintenance but also adds total quality control and total employee involvement as key issues. Traditionally, the maintenance department has been expected to repair broken equipment, doing preventive maintenance when there is time. If a piece of equipment is down, it is worked around—employees stay busy building piles of inventory. With JIT, the flow of product is interrupted if equipment breaks down or maintenance does preventive work without scheduling with manufacturing first. The entire line stops until it can flow again. Employees stop working.

> JIT requires a different approach to equipment maintenance—one that is proactive, involving quality and people.

The fourteen benefits of TPM will be described in more detail after the discussion on setup reduction.

Reducing Setup Time to Increase Flexibility by Involving the Experts

The setup-reduction process that I will describe was begun by my colleague, Ed Hay. After studying setup reduction in Japan, Ed "westernized" the process, and we have developed it into a proven process that works every time in any company. I have found that when this process is followed carefully, a 75 percent reduction in setup time can be achieved with little or no cost. Then the process can be repeated to continue reduction of setup time.

In this process, setup reduction focuses on simplifying the setup—in other words, taking out the waste. Equipment downtime is targeted for elimination. Setup time is defined as the elapsed time from the last good piece in a batch of one product or variation to the first good piece in the next batch.

The biggest challenge is getting people to believe that a 75 percent reduction is possible. In many companies, changing the mindset of management and shop floor workers involved in setup about the possibility of reducing setup time seems to be much more difficult than actually effecting the reduction.

Seeing Is Believing

One of our clients had made impressive progress with flow and Drumbeat. Because the company has a large number of products—approximately

11,000—setup time became a critical issue in efforts to reduce the batch sizes and improve flow. However, the plant manager did not think that there was much opportunity to reduce setup times. After all, his setup people were the best and had years of experience. They knew the machines better than anyone, and engineering support was good. In fact, engineers had been asked to look for ways to reduce setup times and had found few if any.

We convinced the plant manager to have the manufacturing and engineering managers spend some time in manufacturing exploring setup reduction opportunities. All we asked was that they observe a typical setup with us. Finally, they agreed to join us. We picked a typical setup on a Matsuura Milling machine. As we watched the setup, the manufacturing manager and engineering manager were surprised. They couldn't believe how long it took to complete the setup.

They immediately decided to launch a setup reduction effort with three types of equipment—the Matsuura Miller, a Weldon grinder, and a Nakamura lathe.

The documented setup on the Matsuura was 163 minutes. The team has been very successful in reducing setup time by:

- Using the full tool carousel, eliminating the need to take all tools out for each setup
- Eliminating the wait for first piece inspection

The other two teams used a combination of these ideas and added other innovative changes, improving tools and fixtures and instituting better organization, to attain success. Several months later, they have once again proven that 75 percent reduction is a reasonable, attainable goal.

If everyone begins to challenge the way setups are done, you will be off to an excellent start.

Establishing the Right Objective for Setup Reduction

Setup reduction is not a people-reduction program, nor should it focus on reducing costs. The purpose is also not to produce more product, unless the company is completely out of capacity and must create more capacity to satisfy customer needs. The goal of setup reduction is to create more flexibility in manufacturing to build what customers need, when they need it, by reducing the batch size. The ultimate goal is to be able to build any batch size economically, instead of being forced to build more than is needed just to justify the cost of a setup.

By reducing setup times, then reinvesting the time saved into more frequent setups, you will drive batch sizes down so that, instead of building three months' worth of product, one month's worth can be produced; instead of a month's worth, a week's worth can be produced; even in the best of circumstances, instead of a week's worth, a day's worth can be produced.

Let's take a look at a common situation in manufacturing. It takes one hour to set up for a particular job that must run seven hours to build one hundred pieces, according to an EOQ calculation. One hundred pieces is one month's sales. Four different products each sell twenty-five per week. A month's worth of each of these products is produced in four days; the other sixteen days each month the machines are running a host of other products.

The company sells only twenty-five pieces each week, so seventy-five of the one hundred pieces produced are put in stock. What are the consequences?

- If there is a quality problem, one hundred must be reworked or scrapped.
- It costs money to store these extra pieces.
- Flexibility to supply exactly what the customer needs is limited. Resources have been used for product not needed by the customer, preventing use of resources on some products the customer does need.
- If there is an engineering change, the customer must wait at least a month before the change can be implemented, and parts already built are reworked or scrapped.
- If schedules change, response time is slow.

If the setup time can be reduced by 75 percent to fifteen minutes, the savings in setup time can now be reinvested into more setups and twenty-five pieces—a week's worth—can be built in each batch. This point is illustrated in Figure 9-1.

The total setup time remains one hour, but one week's worth of each of four different parts can now be produced in an eight-hour shift. The other four days of the week other product can be made; next week the cycle begins anew. Resources are now used for what is needed now. Response time to the customer for a new order or a change is also drastically reduced. If there is a quality problem, feedback is immediate, so rework and/or scrap are reduced by 75 percent as well.

Quality improves in another way because "practice makes perfect." As setups are done more often, quality of the setup improves. An operator may not remember a setup if he or she does it only once a month, but a weekly setup becomes routine; a setup done four times a day should be smoother than one done once a day. Efficiency may also improve as a

Figure 9-1. Reducing setup to provide flexibility.

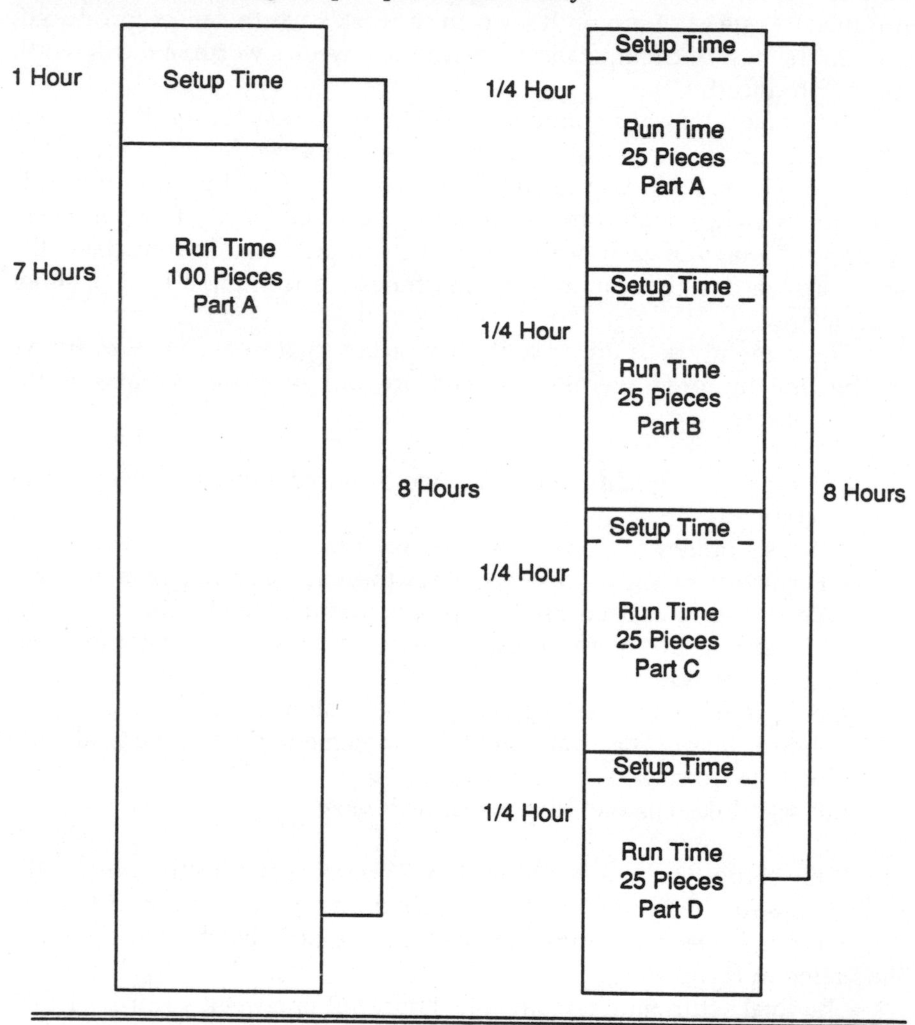

result of practice, reducing cost in some cases. However, this can never be the main impetus for setup reduction.

Establishing the Experts for Setup Reduction

Setup reduction is not an engineering project. This may sound strange, since engineers are the technical experts for the equipment. But that's exactly the reason why this is not an engineering project: Engineers immediately dive into the equipment to explore technical improvements, which are not usually low-cost solutions.

Successful setup reduction requires a team approach. The makeup of the team should be primarily setup people and operators, with technical and management support as required. Successful setup reduction teams that we have worked with always have at least one engineer on the team to provide technical support, but the engineer cannot be allowed to control the effort. The teams can also have members of management, planning and scheduling, or industrial engineering. A guide—someone with experience executing this process—is also important; this could be a consultant at first, until the company develops in-house experience and expertise.

The message should be crystal clear:

> The experts are those who do the setup day in and day out.

The workers performing the setup and the operators (this may be different workers or the same ones) should make up the majority of the team. They are able to focus on downtime and simple improvements. Many of these workers have been storing ideas for years, just waiting for someone to ask for suggestions about improvement.

These teams are empowered to implement changes. They need to be able to influence groups outside of their own team as necessary to get the job done. Some companies have established a small budget for each team that can be used without the team having to fill out requisitions or purchase orders for little items. Teams with such authority are generally much more careful about how money is spent than managers who have sign-off responsibility and accountability.

The Seven Steps for Achieving 75 Percent Setup Time Reduction

The seven steps for achieving a 75 percent reduction in setup time should be followed in the order described. Each step is essential for success.

Step 1: Get Organized

Management support must be obtained during this step. Management should understand the what, why, and who of setup reduction and agree to the objectives as discussed. Usually, a steering committee is established to pick the equipment or process and to establish teams. I recommend no more than three teams to begin with. Three teams provide an opportunity to get results in three different areas, providing significant impact without significant risk of overloading support functions that may be called upon to help. Also, three teams is the maximum number that can be trained together, utilizing your guide or internal resource well.

Some companies designate a coordinator for the setup reduction effort in order to provide continuity between the teams and future efforts. This internal resource should eventually become the guide.

Team members should be identified by the steering committee or the management leading this effort and each member provided the opportunity to buy into the process. In addition, communication sessions should be held with the entire work force to explain the what, why, and who of the effort.

Step 2: Train the Team

Training for the entire team membership is essential to get off to a good start. Most teams devote two to three days to team training. Training is also a must for each team facilitator.

Since this is a focused task team, the tendency will be to get on with the technical issues rather than to spend time on interpersonal issues. However, setup reduction teams who have taken the time for training have been significantly more successful than those that have jumped to the technical issues right away. In fact, my consulting firm no longer accepts the 75 percent reduction challenge if a company declines the training effort.

Step 3: Videotape the Setup

The next step is to videotape the setup. Video equipment is an essential tool for any improvement effort and should include a visible time with seconds noted. If your company does not own such equipment, it should be purchased or rented. Generally, the steering committee chooses the setup as it chooses the team, but in other cases the team chooses the specific setup. Pick a setup that is executed often and is representative of other setups in the area. A more difficult setup will usually cover more issues and may be a good choice for maximum benefit. This video should belong to the team, and outside review of the video should be done only with the team's permission. Some people fear that the video will be used to find fault or do time studies on the operation. They must be confident that this is strictly a tool for them to use to improve the setup.

Step 4. Document and Analyze the Video

The team facilitator and the person executing the setup in the video should document the video with a flow analysis, or step-by-step analysis. The steps can then be categorized into at least four classifications: internal/external, adjustments, clamping, and problems. The team may add other classifications to this list if it makes sense to do so. The time to complete each step should be included.

1. *Internal/external* focuses on equipment downtime (see Figure 9-2). The idea is to keep the equipment running as much as possible. Look for those activities that can be executed while the equipment is running—not when it is stopped. The effort is to get ready for the next setup while the equipment is running.

For example, many teams have been able to reduce setup time significantly by presetting tools rather than setting the tools while the equipment

Figure 9-2. Focusing on machine downtime to reduce setup time.

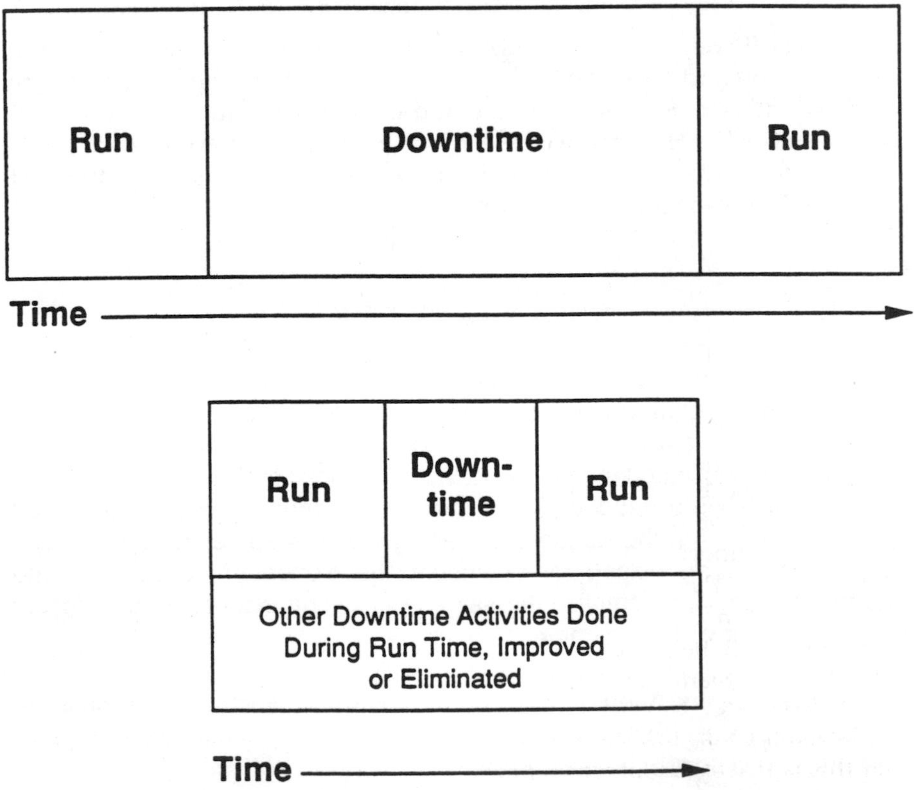

is down. In some cases, it is possible for the operator to do this while the equipment is running; in other cases, people in the tool crib have been given the task of presetting tools.

Reducing Downtime

Back to the story about the Matsuura setup reduction effort. Traditionally, only one set of tools had been allowed at the machine at a time in order to keep control of the number of tools outside the tool crib. Therefore, after a particular job was complete, the operator had to return the tools to the crib and ask for the next set for the next job. The tools were put into the holder and the operator returned to the machine to complete the setup. All this was done while the machine waited.

The team began to look for setup reduction opportunities and realized that many tools were used for many jobs. If some holder spaces could be dedicated to certain commonly used tools, the tools would have to be replaced only when dull—not between each job. Identifying tools and spaces and changing some programs led to a 26 percent savings on setup time.

2. *Adjustments,* the next category in the analysis, occur when the first piece is made and then checked to determine proper adjustments before the next one is run. This is the trial-and-error approach. The question to ask constantly is: How can we be certain that the first piece is good every time? Adjustments are part of setup time—the clock runs until the first good piece is made after all adjustments.

Speeding Up Adjustments

The Nakamura team mentioned earlier in this chapter made a 35 percent improvement in setup time just by paying attention to adjustments.

In the past, team members had depended on the quality department and some sophisticated measuring devices to check their first pieces to see if adjustments were needed. Since this quality department was a shared resource away from the Nakamura, waiting was inevitable. To reduce such waiting, the team, helped by engineers, developed some simple hand gauges that could accomplish the same check. The inspection now occurs right in the area by the operator.

3. *Clamping* is another area of opportunity. Focus on eliminating threads of all sorts—they take lots of time plus extra tools. Try to replace threads with clamps whenever possible.

Reducing Use of Threads

One HP division built many different small printed circuit boards, using an automatic component inserter. Each PCB required a plate the exact shape of the board to hold it in place while components were added. This plate had to be secured with threaded bolts each time.

The team made a plate with several different PCB shapes and sizes, which eliminated some changes of the plate completely. And when the plate did change, it was changed to clamp into place—no more threads.

4. *Problems* constitute the fourth category—all kinds of problems. They may be as simple as having to look for a lost tool or as complex as a process problem that is preventing good quality on a consistent basis.

Solving "Little" Problems Can Yield Big Results

Problems can encompass a number of different kinds of issues. The Weldon team mentioned earlier in this chapter did a good job of looking for many smaller items that added up to significant savings.

They used Allen screws for setup; the procedure had always been to go to the tool room or maintenance to get them. The team achieved a 6 percent savings just by stocking these screws in the work area. Another 5 percent was gained just by organizing the tool cabinet to expedite tool and fixture retrieval. A 3.3 percent was achieved just by moving a Dumore grinder into the area to eliminate the need for the operator to move to a different area for just one operation. A 14.3 percent reduction was achieved just by paying attention to all those "little" problems that get in the way!

Step 5. Identify the 75 Percent

Through review of the video and a general brainstorming session, the team will be able to generate ideas about how to reduce the setup time. Members should review the video carefully and slowly, constantly stopping it to note opportunities. The brainstorming session should focus on listing all the complications that get in the way of doing a better job. A value-added analysis should be done to indicate nonvalue-added steps.

The next step is to rank-order items to work on, focusing on the most time-consuming steps first, trying to find a total 75 percent reduction. Make certain that the team develops a list of "doable" items that can yield a 75 percent reduction. Complete the list before beginning to implement any item on it.

Step 6. Implement the Changes

Implementing the changes to achieve the 75 percent setup-time reduction will take a long time. It may involve technical support, management, or even equipment suppliers. The team must develop a realistic implementation plan and be willing to revise the plan as it goes along. The key here is that the team develop and monitor the plan to ensure that the changes are made.

Step 7. Videotape the Results and Expand the Setup Reduction Effort

The video can be used during the process of getting to the 75 percent at any point that the team feels it would be useful. When the team thinks it has achieved its goal, the new setup should be videoed. Not only will the tape serve as a training aid for future teams but it gives the team a real sense of accomplishment to be able to document success in reducing the setup.

The next step is to implement the changes made in achieving the 75 percent reduction everywhere possible. This may mean making changes on other setups in the same area or expanding to other areas where there are similar setups. Many companies use similar equipment, tools, and line configurations in many different areas. Make sure that you take full advantage of what has been learned. The goal is to reduce every setup by 75 percent. Do not stop until every setup on every machine has been reduced 75 percent. Then begin again through the process, striving for another 75 percent. Then begin again through the process, striving for yet another 75 percent.

Eventually, additional reduction in setup time will involve capital investment. The cost and benefit must be evaluated carefully to decide whether it is worthwhile to spend money to improve the setup beyond the low cost/no cost improvements.

How to Reduce Setup Time in High-Tech and Nonautomated Environments

High-tech companies have opportunities in setup reduction that have gone unnoticed for years. Since there is high variation of products in most of these companies, setup is an issue even though the setup may be measured in minutes instead of hours.

High-Tech Setup Time Reduction

One of the AGFA Compugraphic areas was responsible for testing all printed circuit boards that were built, using electronic testing equipment involving different software to be loaded for every different board. This setup was in the fifteen-minute range. Most people would not worry about such a short setup, but fifteen minutes was too long here. Every batch was different. The team applied the seven steps and reduced the setup to the three-minute range. Some of the improvements involved moving the test fixtures closer, making sure that in every case the example board was good and updated, and storing and loading the software in a more efficient manner.

Setup reduction is applicable even in nonautomated environments. Setup includes any activities necessary to change from one variation of product to the next.

Reducing Setup Time in Nonautomated Settings

At one of the high-volume electro-surgical pencil lines at Valleylab, the line must be completely cleared before a new product is begun in accordance with FDA requirements. This means that all parts and tools for a product completed are removed from the line before new parts and tools for the next product to be built can be arranged on the line. This procedure is designed to ensure that parts are not mixed up from one product to the next. Setup on this line may involve only getting different parts and a few tools. Some rearrangement of work stations and equipment may occur in order to balance the flow of product. However, when a company is building thousands of pencils each shift, minutes count! This setup should be attacked and reduced to the absolute minimum possible time from the last good part of one order to the first good part of the next.

Stabilizing the Process With Total Productive Maintenance

Total Productive Maintenance (TPM) is required to implement JIT successfully. Having equipment and tools that are always in peak operating condition can remove many unexpected variables from the process and provide more stability. Stability leads to predictability and improved quality, two key principles of JIT.

At first glance, it may be difficult to see how maintenance and the maintenance department is affected by JIT. I think of Joe, the maintenance manager of a manufacturing company. Joe had worked his way up through the ranks, beginning as an assistant many years ago. Since maintenance people are expected to know a little about everything, most of our very best maintenance people graduated from the school of hard knocks. Joe had done just that—he could repair most anything in the shop. His world was one of little stress most of the time—spare parts were expensed, so no one bothered him about keeping track of them, and Joe's people knew through experience what to order.

Manufacturing would call Joe when it had a problem, and Joe would get to it when he could. When there was time, Joe would try to do a little preventive maintenance, just showing up in manufacturing without scheduling. Manufacturing was so happy that Joe was there to do a PM that people allowed Joe to use the equipment and worked around him. The maintenance department tended to be the forgotten one down in the corner of the building.

But JIT requires a constant flow of product with 100 percent quality and does not allow queues to build up. The Drumbeat must be met every day. Suddenly, with the introduction of JIT, Joe gets a completely new spotlight on his department. The requirements for maintenance to support a JIT environment are very different. Now Joe gets the call stating that a piece of equipment is down. The line stops until the equipment is repaired. The line can't wait "until I can get to it." The line can't wait until Joe orders the parts. The line can't "work around" the problem. And the workers on the line are asking why the equipment went down in the first place. The pressure could be unbearable for Joe.

So the maintenance department needs to understand and prepare for new requirements from JIT. Total Productive Maintenance differs from traditional breakdown and preventive maintenance in many areas.

The Fourteen Benefits of TPM

1. In traditional manufacturing, the maintenance people "fix" things. Under TPM, maintenance is part of the business strategy.
2. In traditional manufacturing, maintenance is reactive—"*fix it* when it's broken." Under TPM, maintenance is proactive, to eliminate breakdowns.
3. In traditional manufacturing, there is accidental cost with no measurements. Under TPM, there are planned and controlled expenses with specific measures.
4. In traditional manufacturing, maintenance is expected to maintain new equipment when it arrives on the line. Under TPM, maintenance is a part of the equipment selection and design.
5. In traditional manufacturing, there are maintenance inventories; the goal is to have plenty of everything and not to run out. Under TPM, the goal is to have the right parts at the right time, carefully planned.
6. In traditional manufacturing, only maintenance specialists are expected to do maintenance. Under TPM, maintenance is everybody's job, and the maintenance department should be trainers and coordinators. Specialists should take care of only those tasks too difficult for anyone else.

7. In traditional manufacturing, ideas for the equipment come from engineering, and maintenance just executes them. Under TPM, maintenance generates and implements ideas.
8. In traditional manufacturing, the maintenance department is made up of responding individuals. Under TPM, the maintenance department is a "proactive" thinking team.
9. In traditional manufacturing, a few rejects are okay. Under TPM, 100 percent quality is required 100 percent of the time.
10. In traditional manufacturing, quality is the quality department's job. Under TPM, quality is everyone's job—including maintenance's.
11. In traditional manufacturing, if equipment is down, manufacturing works around it. Under TPM, manufacturing stops until the line can flow again.
12. In traditional manufacturing, maintenance gets to preventive maintenance when it can. Under TPM, PM is carefully planned and scheduled with manufacturing to ensure that it happens regularly without disrupting the flow.
13. In traditional manufacturing, maintenance is a service department to manufacturing—call 'em when you need 'em. Under TPM, maintenance is a vital part of the manufacturing team.
14. In traditional manufacturing, there is breakdown and preventive maintenance. Under TPM, there is the addition of Total Quality Control and Total Employee Involvement to breakdown and preventive maintenance to create Total Productive Maintenance.

Beginning a Successful Maintenance Team

Forming a team in the maintenance department is no different from forming any other team. It is essential to begin with the education phase discussed in Chapter 3. In order to capture ideas and improvements, the team should meet at least once a week, away from day-to-day activities. Since maintenance people often work independently, it may also be worthwhile for them to be on other teams in manufacturing.

Don't forget maintenance when beginning a JIT effort. It can greatly affect the success of JIT and the speed at which you progress with the implementation. The maintenance department needs time to plan and get on board to develop a Total Productive Maintenance mindset.

Setup Reduction and Total Productive Maintenance are two vital pieces of any successful JIT implementation and should not be ignored, regardless of the type of business you are in.

Chapter 10

Applying JIT Beyond Manufacturing: Making It Work in Administrative and Support Functions

JIT is not just for manufacturing anymore. In fact, it never was.

Many companies begin a Just-In-Time implementation in manufacturing, achieving relatively quick results. However, some of the largest gains can actually be achieved outside manufacturing. In most companies, greater than 50 percent of product cost is overhead. By starting at the heart of the company—manufacturing—a pull is created for support functions to do their work differently in order to provide what is needed by the new way of manufacturing. The JIT mindset and principles can benefit *everyone* in the company.

In this chapter, I discuss how to define these nonmanufacturing functions in such a way that the same principles discussed earlier can be applied to them. Next, I outline three ways that JIT can help these groups implement that continuous improvement mindset and summarize how the JIT principles discussed earlier work in nonmanufacturing environments. Finally, I describe the nine key steps for implementing JIT in support and administrative functions.

Defining the Process, Customer, and Supplier for Nonmanufacturing Functions

Every function in your organization has a process. It is easy to define a process in manufacturing; just follow the product as it gets built, and that is the process. In a nonmanufacturing area you need to define the product being "built." This could be a customer order, a check to a supplier, a letter, or an engineering documentation package.

Step 1: Define the Process

Once the "product" is clearly defined, you can simply follow the product through the process, noting the process steps. For instance, to produce a check in accounts payable for a supplier, the process may include steps displayed in Figure 10-1.

Of course, this is the process if everything is correct. Many rework loops exist if problems are found. In any case, the process for producing a check has been outlined.

Step 2: Define the Customer

Once the product and the scope of the process have been defined, you can define the customer for the particular product. The customer is the next person in line to receive the product. For instance, the customers for an engineering documentation package would be manufacturing and materials, since manufacturing needs the documentation to know how to build the product and materials needs the documentation to order parts to build the product.

Step 3: Define the Supplier

When the product and the process have been defined clearly, the supplier can be determined as well. Often, several suppliers can be identified. Let's take a stockroom (or stock area) for production materials as an example. If the parts are inspected, incoming inspection is a supplier; the supplier company is a supplier of parts (if they do not get inspected); and the documentation group supplies information on storage requirements and shelf life. Engineering may supply information to the stock area on how the part is used, determining where to store the parts.

Now we can begin to look at any process, manufacturing or nonmanufacturing, in the same way. The key is to help all of your people involved in nonmanufacturing processes think of their function as a process with a product, customer, and supplier.

My colleague Joseph Bonito has outlined the three reasons why JIT is such a powerful tool in nonmanufacturing improvement efforts.* They are as follows:

*1990 article, "Beyond Manufacturing: Putting JIT to Work in the Office," in The Journal of Manufacturing: Issues, Options and Strategies in the Automated Workplace, Vol. 1, No. 4, Winter. 1990.

Figure 10-1. Steps in the process of producing a check for a supplier.

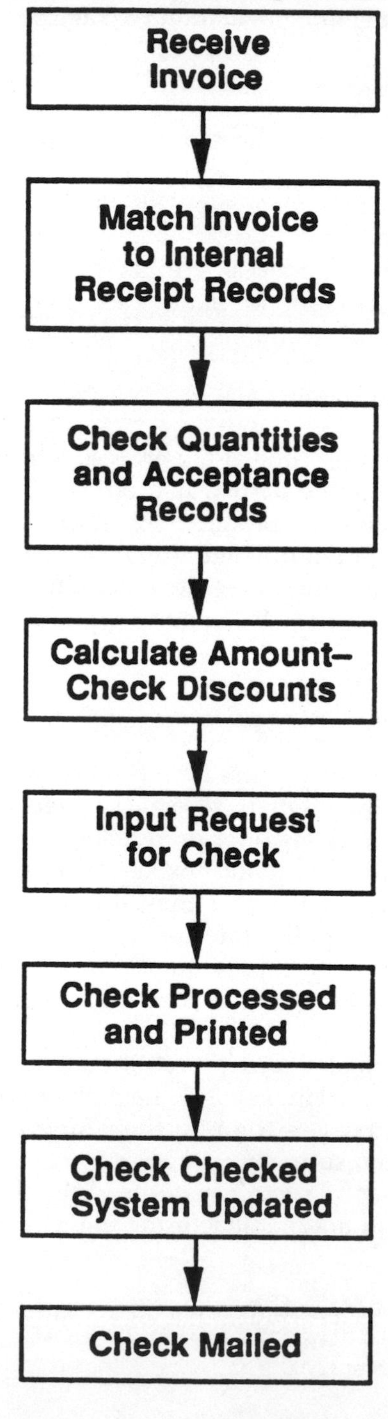

Three Ways JIT Helps Nonmanufacturing Functions With Continuous Improvement

1. *JIT clarifies opportunities.* Following the process-improvement principles brings to light many opportunities to make things better in nonmanufacturing areas. By putting clear objectives, goals, and measurements into place, your people will feel that they, too, can move ahead. I have found that many people in administrative functions just come in, do the job, and leave very tired. Many administrative jobs today do not offer the reward of completion of product or task; it is more like being on a treadmill than crossing the finish line in a race.

JIT can clarify opportunities for improvement by baselining the process and searching out activities that do not add value.

2. *JIT empowers people to make improvements.* By encouraging everybody to question processes and procedures that do not make sense, JIT helps identify opportunities for improvement. But JIT goes beyond that: People can also *implement* improvements, giving nonmanufacturing people the chance to work smarter, not harder. JIT gives people the power to identify and implement changes that make sense.

3. *JIT connects nonmanufacturing processes with satisfying the customer.* The ultimate success of your company rests on the relationship with the external customer, but most work produced goes to an internal customer—the next employee receiving the output. JIT forces workers to analyze the work being done from the perspective of both the internal and external customer.

If the nonmanufacturing process flow does not meet internal and external customer needs, there is an opportunity for improvement. JIT is a customer-oriented improvement process that can correct administrative and support processes with customer satisfaction.

Putting Key JIT Practices to Work in Nonmanufacturing Functions

The foundation building blocks also apply in nonmanufacturing functions.

1. *People.* Surely the People principle applies to 100 percent of the employees 100 percent of the time. As people get more involved in process improvement, they have more ownership for that process and become committed to continuously improving—whatever the task. Even the most repetitive administrative job can be improved if everyone gets that continuous improvement mindset.

2. *Quality.* Clearly defining an administrative or support function can help establish clear quality goals and measurements. Quality problems

abound in nonmanufacturing areas, but in most cases, these problems lack focus. How many times does a letter get reworked? What about all the inspection points in a customer-order processing cycle? Checks and balances run rampant in nonmanufacturing as a way of life as part of the process. As problems are clearly identified and solved, these nonvalue-added checks and balances can be eliminated.

One of the best ways to get going with an improvement effort is to define clearly quality measurements and to report against these measurements. Visible, easy-to-understand measurements are essential. Every nonmanufacturing function should define quality measurements that make sense for its process and post these measurements regularly.

3. *Drumbeat.* Drumbeat is also applicable in nonmanufacturing processes. Synchronization with the customer is surely a possibility once the customer and customer's requirements have been defined. Linear usage of capacity may be more difficult in nonmanufacturing areas because of the short lead time between customer demand and delivery and the lack of a forecast. Think of customer service people: How do they linearize their flow of work when they have three phone rings to notify them of the demand? In this case, the way work is designed to meet this short demand cycle may be the key to linearizing the use of capacity. Drumbeat provides the opportunity to connect work into a flow instead of being a series of discrete events with in-baskets between each one.

4. *Flow.* The product flow through different processes can be mapped out for a nonmanufacturing product using the same methodology described in Chapter 6 for a discrete product. Figure 10-2 shows the original product flow of parts as they are kitted in a stockroom of a build-to-customer-spec electronics company.

After the stockroom team analyzes the steps in the product flow and maps out the layout, improvements become obvious. Figure 10-3 shows the new layout and product flow.

With this improved flow, mileage that parts (and people) travel is reduced 53 percent, while output increases 240 percent with the same number of people.

Now, your question may be, "Why work on flow in a stockroom? Aren't we going to eliminate stockrooms anyway?"

You are right. The long-term goal is to have 100 percent quality parts go from the dock to manufacturing. However, in a custom job-shop environment, where certain parts are purchased specifically for a certain job, there may always be a need to do some kitting. In the meantime, why not improve the flow by getting the stockroom team involved so they can help figure out ways to improve even more?

5. *Value added:* The principle of value added surely applies in nonmanufacturing areas. Many steps in these processes do not add value to the product and contribute to meeting the customer's requirements.

Figure 10-2. Original flow of parts to be kitted.

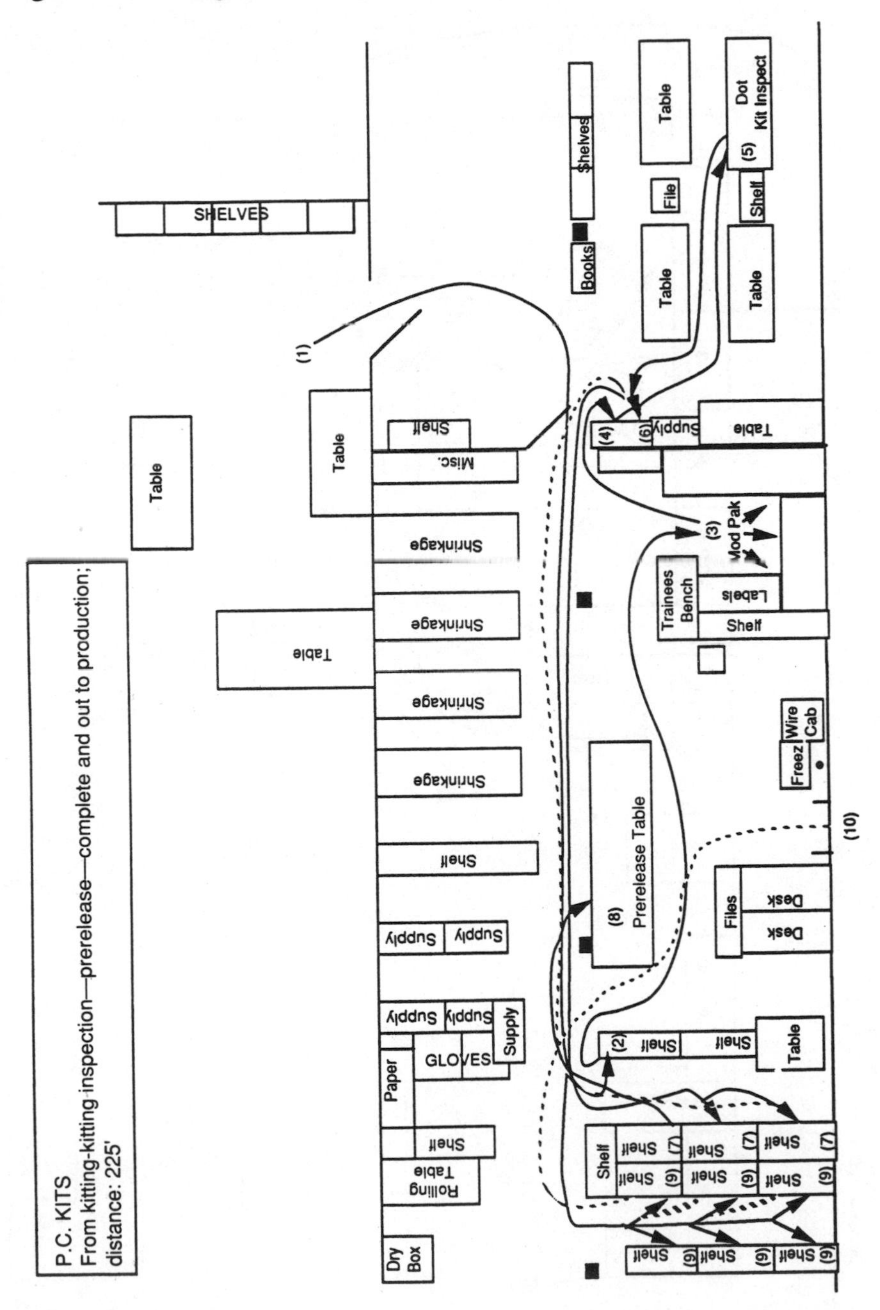

Figure 10-3. The new layout for better flow.

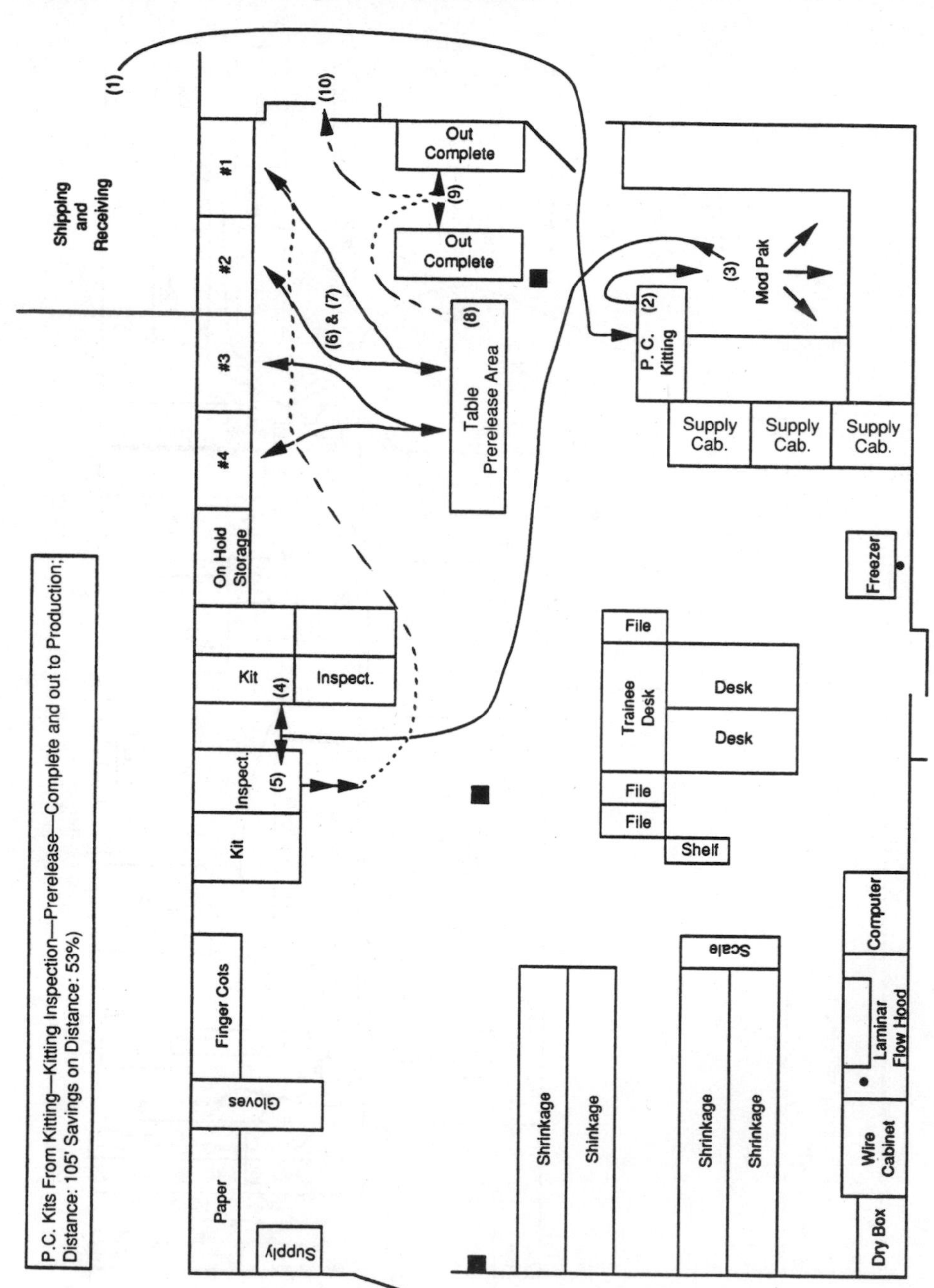

Exhibit 10-1 at the end of the chapter is an accounts-payable team value-added analysis and resultant improvement for the expense report process at Thomas & Betts, headquartered in Bridgewater, New Jersey. With 4,700 employees worldwide, the company produces automotive products and electrical and electronic products. This team used the value-added analysis as the basis for improvement. Notice the outstanding results!

6. *Centers.* As you recall, centers combine unlike processes to improve the product flow.

Recently, an incoming inspection team—those who inspect parts from the supplier—had a cycle time that was unacceptable to its customer (manufacturing). Parts, on average, stayed in incoming two weeks. Since it was not possible to eliminate immediately this inspection step, the team decided to look at the flow of product through the area. The team documented the process flow as shown in Figure 10-4 and found that improvement ideas were plentiful.

The group was organized by type of test equipment, with all docu-

Figure 10-4. A flow of incoming inspection.

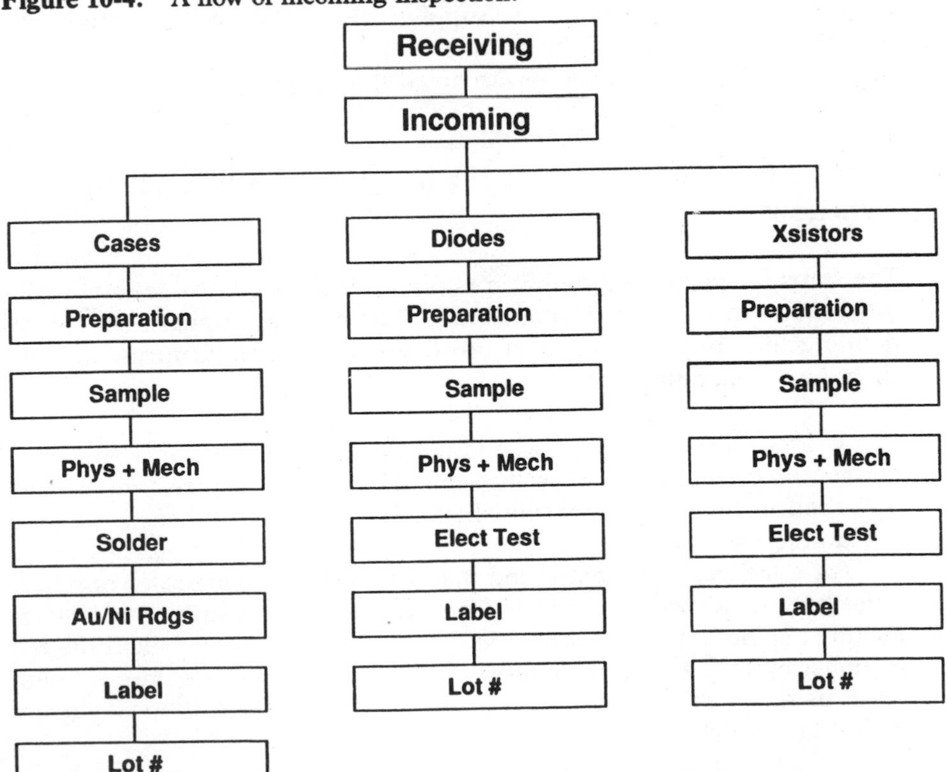

mentation files in one part of the area. Processes were grouped together (Figure 10-5).

As group members began to look at the types of parts they inspect, it became clear that they had four families of parts and that each utilized particular equipment and files. The team created product centers, or cells, in incoming inspection (Figure 10-6). Cycle time plummeted within one month. One day, the team ran out of work, and members were able to help receiving.

By applying the center principle, flow mileage was reduced 80 percent, cycle time was reduced 77 percent, and output was increased 26 percent. Now this team can focus on eliminating incoming inspection over time.

7. *Demand pull.* Demand pull—building product only when the customer needs it—also applies to nonmanufacturing functions. Demand pull provides a simple, visible system for applying resources where they are needed most to meet customer requirements.

One of the best demand-pull implementations I have seen was an accounts payable process for producing checks. Instead of in-baskets and files, demand signals were set up to prevent large queues of in-process checks. The cycle time was drastically reduced, giving the company an opportunity to take many more supplier discounts. The capacity of the department increased to the point that it could handle the increase in number of shipments/invoices with no additional people.

Nine Steps for Implementing JIT in Administrative and Support Functions

The steps for implementing JIT in nonmanufacturing areas are not really different from those in manufacturing. There is an emphasis on clearly defining the process, since most processes in nonmanufacturing are less clear than in manufacturing.

1. *Pick a function for focus.* The vision discussed in Chapter 1 applies to nonmanufacturing areas as well as manufacturing. Business imperatives described in the company vision should be considered when choosing a function to focus on.

For instance, one client found that it was losing some sales opportunities because it was not producing bids fast enough. Business imperatives included some healthy growth objectives, so the company picked the bid response cycle as a function to focus on. Through its process-improvement effort using JIT principles, the company is experiencing a 40 to 50 percent cycle-time reduction in bid responses.

2. *Define the product, process, customer, supplier.* Once a focus is deter-

Figure 10-5. The original flows.

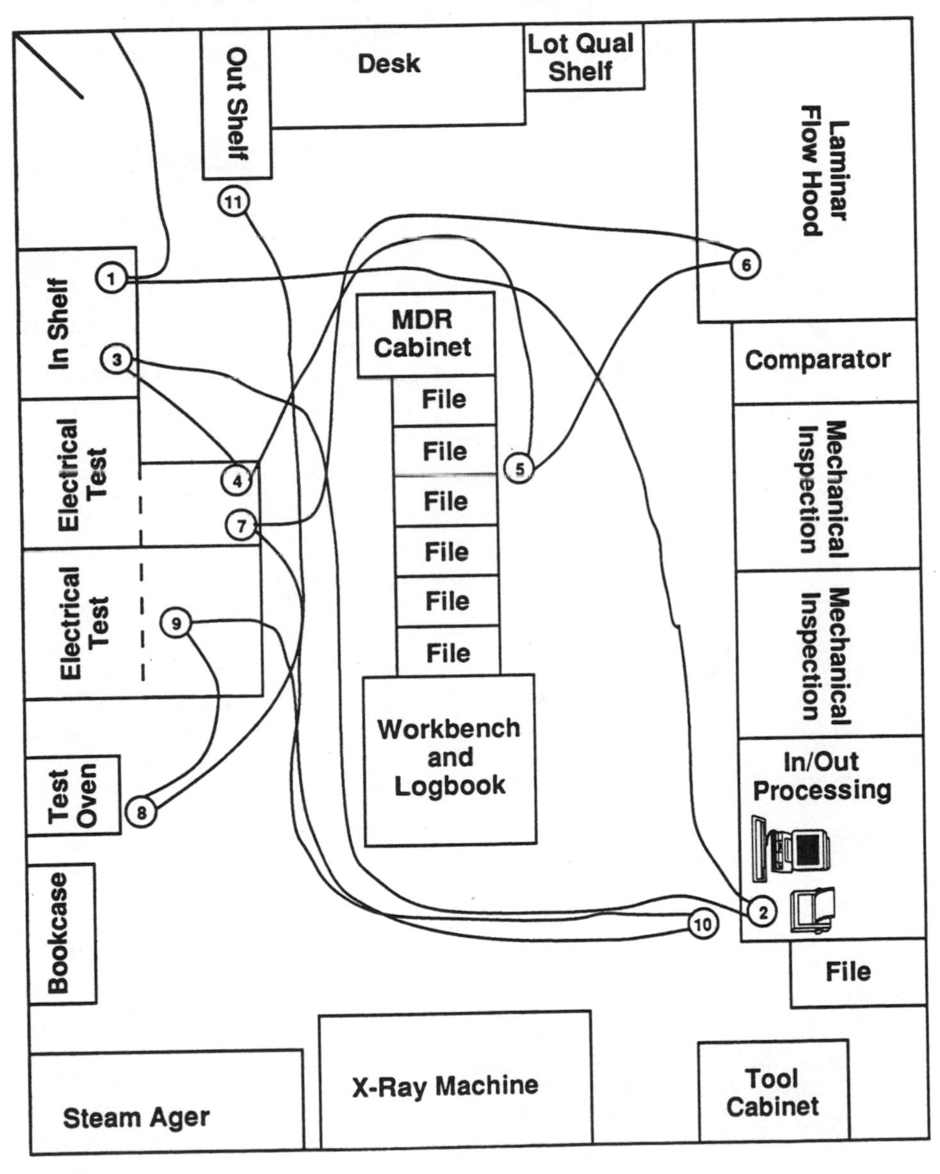

Figure 10-6. The improved flow using the center principle.

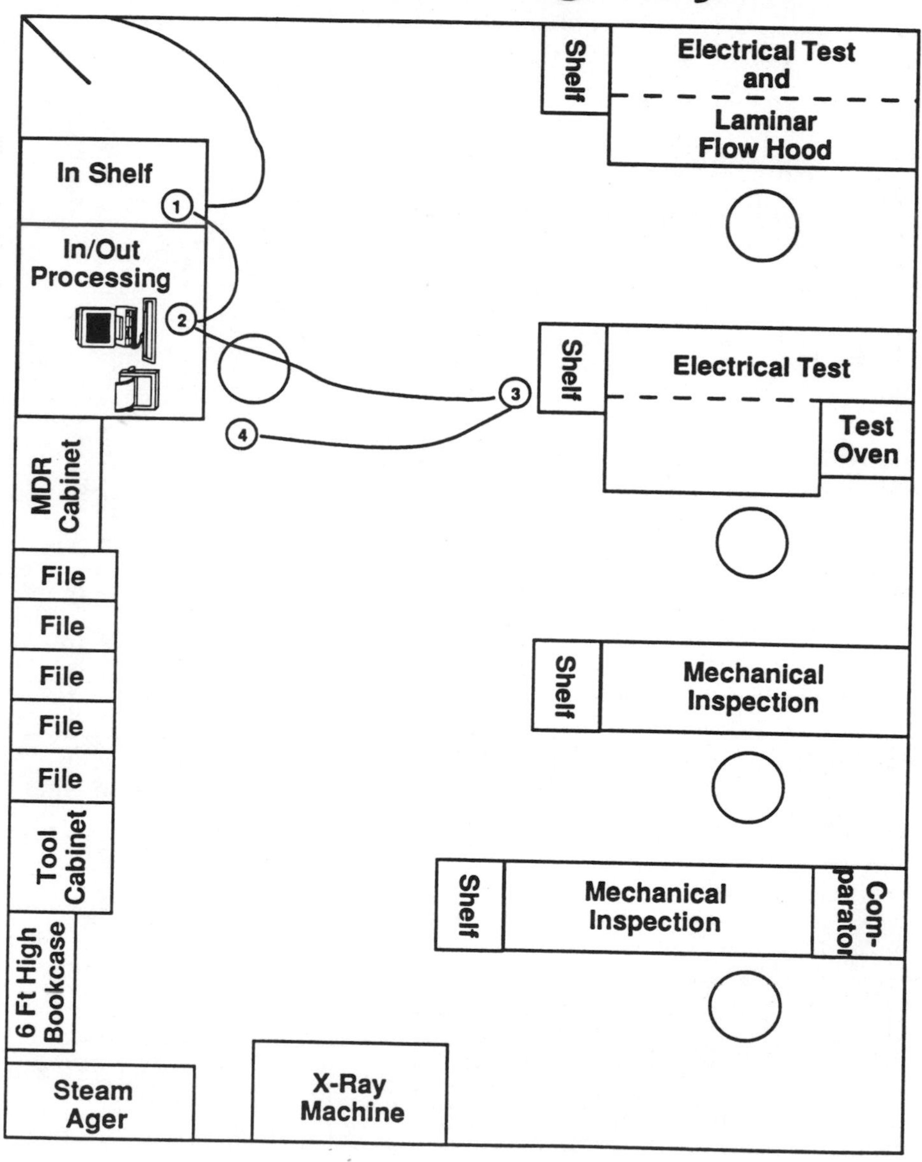

mined according to business imperatives, it is necessary to define the product, process, customer, and supplier, as mentioned earlier in this chapter.

3. *Perform global value-added analysis.* This step ensures that the function picked is value added. Many administrative activities are not required to meet internal or external customer requirements. They have simply been institutionalized at some point for some reason and never evaluated again. For instance, maybe instead of creating an accounts-payable process to handle more volume, a company should reduce the number of suppliers, negotiate blanket orders, or link to suppliers electronically.

4. *Define customer requirements.* Many times we assume that we understand what the customer requires. However, when asked, the customer may give a response different from those assumptions. A customer-requirements survey should always be done to ensure a complete understanding of customer requirements. One customer service department assumed that its customer, the credit department, wanted orders once a day in a batch so that all interruptions and questions could be handled only one time a day. However, when the customer service department interviewed the credit department, it found that the customer would much prefer a constant flow of orders and information throughout the day!

5. *Perform a detailed value-added analysis.* This value-added analysis is guided by the customer requirements determined in step 4. In other words, determine whether steps in the process are value added from the customer's point of view.

6. *Gather data and baseline.* As additional data are gathered to determine the present process and status—a baseline—some "quick wins" will be apparent. Ensure that there is provision for implementing these obvious improvements as quickly as possible.

Hutchinson Technology, Inc., located in Hutchinson, Minnesota, has done the most complete job I have seen of creating a process for documenting a nonmanufacturing process. The company's focus has been on parts to supply the computer disc-drive business, but the company, which employs about 2,200 people, is branching into other areas.

Reliability Assurance

Two years ago, Hutchinson Technology began moving JIT concepts into the indirect areas of its business, getting groups from maintenance to accounting thinking of the services they provide as "product" and looking for ways to determine customer needs and to meet those needs with quality performance. The program, called "Reliability Assurance," has three phrases. (See Exhibit 10-2 at the end of the chapter). At each phase of the reliability-assurance program, the team doing the self-assessment must meet with a review board made up of "customer" management, "vendor" management, and officers or directors of HTI before it is allowed to proceed to the next phase. Throughout the program, there is an insistence on quantifiable, measurable specifications of each item.

7. *Develop an implementation plan.* The improvement implementation plan should include detailed goals for obtaining the overall objective of the effort and objective measurements for evaluating the plan's success. Milestones detailing who, what, and when should be developed.

8. *Implement.* Allow the people in the function to implement the improvements as they determine them.

9. *Ensure continuous improvement.* Consistent measurement and evaluation of the process will validate improvement efforts. Just as with any other improvement team, do not forget to celebrate the accomplishments of the team.

Some of your greatest potential gains are in administrative and support processes. The JIT mindset and principles apply to every part of your business. I advise that you begin a nonmanufacturing JIT effort early to avoid the assumption that JIT is just for manufacturing.

Exhibit 10-1. A value-added analysis.

Thomas & Betts

Accounts Payable Administrative Improvement Team

Team Charter: Expense Reports

WE ARE A GROUP OF:

1.) Employees
2.) Diverse Personalities
3.) Gathered together to improve A/P Department

REPRESENTING:

Thomas & Betts Accounts Payable Department

TO ACCOMPLISH MAJOR OBJECTIVES:

1.) Work efficiently.
2.) Improve and reduce paper flow.
3.) Define check request procedures.
4.) Show the need for an on-line system.
5.) Improve expense report.

TO ACCOMPLISH WE WILL:

1.) Work together as a team.
2.) Listen to one another and not take suggestions personally.
3.) Go through each part of the expense report.
4.) Get other departments to complete paperwork correctly.

NOT WITHIN OUR SCOPE TO:

1.) Stop doing things (i.e., copies).
2.) Eliminate things that are required.

Thomas & Betts

Accounts Payable Administrative Improvement Team

Team Charter: Expense Reports

ASSESSED BY THE FOLLOWING CRITERIA:

Short Term: 3–5 Months

1.) Reduce time and paperwork by 25%.
2.) Develop job descriptions and procedures.
3.) Get our ideas and recommendations approved by management.

Long Term: 6–9 Months

1.) Get other departments to follow proper procedures.
2.) Save T&B dollars.
3.) Have an on-line system.

Thomas & Betts

Accounts Payable Administrative Improvement Team

Customer Requirements

Quality = Equal to the Expense Report Submitted

Delivery = Pick Up When Ready

Timely = Within a Week After Submission to AP

Thomas & Betts

Accounts Payable Administrative Improvement Team

Value-Added Concept

1. **Define steps or tasks that are necessary to meet customer requirements.**

2. **Add value to the product or service.**

3. **Ultimate Goal: to eliminate nonvalue steps.**

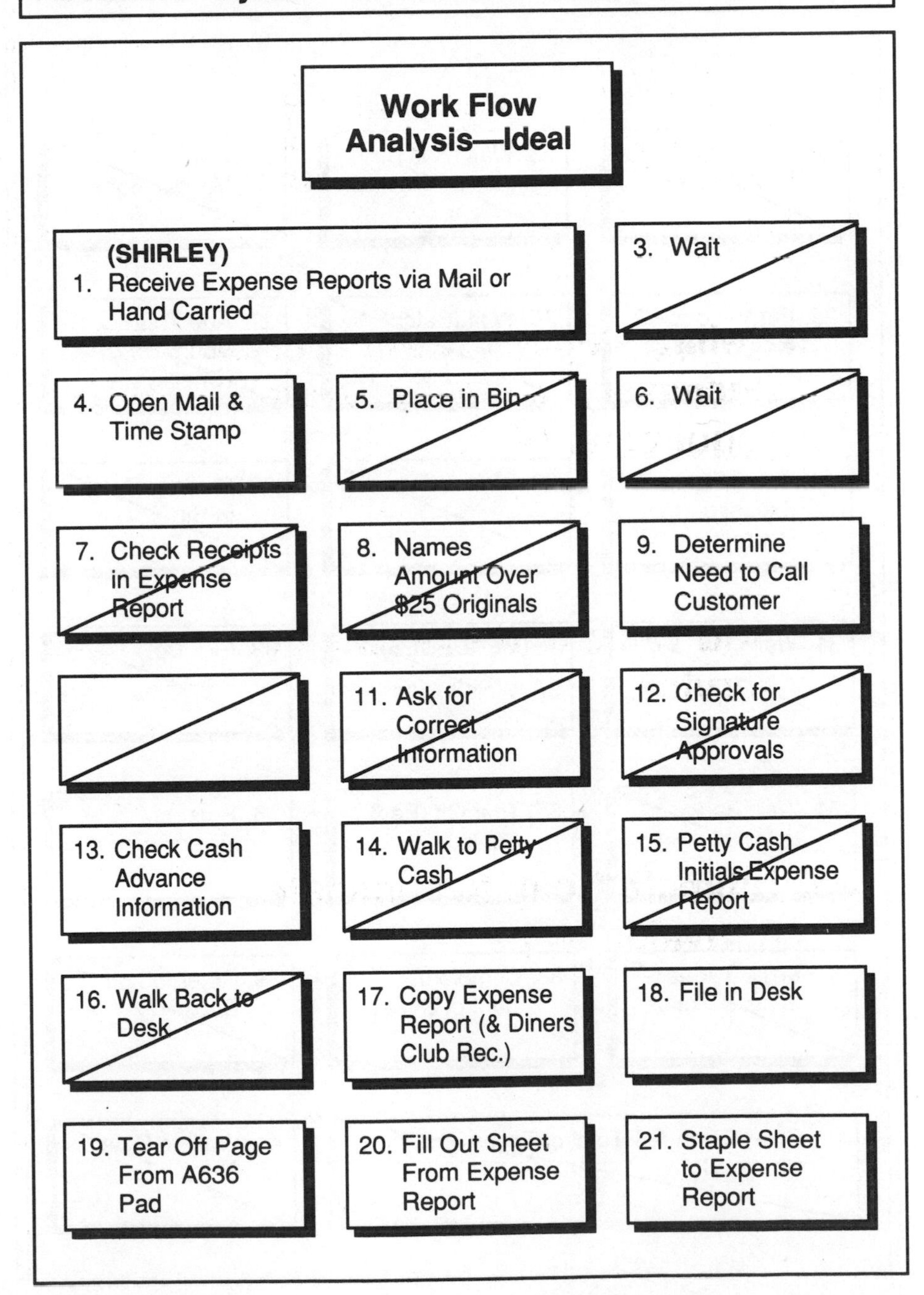
Thomas & Betts
Accounts Payable Administrative Improvement Team
Work Flow Analysis—Ideal
(SHIRLEY)
1. Receive Expense Reports via Mail or Hand Carried
3. Wait
4. Open Mail & Time Stamp
5. Place in Bin
6. Wait
7. Check Receipts in Expense Report
8. Names Amount Over $25 Originals
9. Determine Need to Call Customer
11. Ask for Correct Information
12. Check for Signature Approvals
13. Check Cash Advance Information
14. Walk to Petty Cash
15. Petty Cash Initials Expense Report
16. Walk Back to Desk
17. Copy Expense Report (& Diners Club Rec.)
18. File in Desk
19. Tear Off Page From A636 Pad
20. Fill Out Sheet From Expense Report
21. Staple Sheet to Expense Report

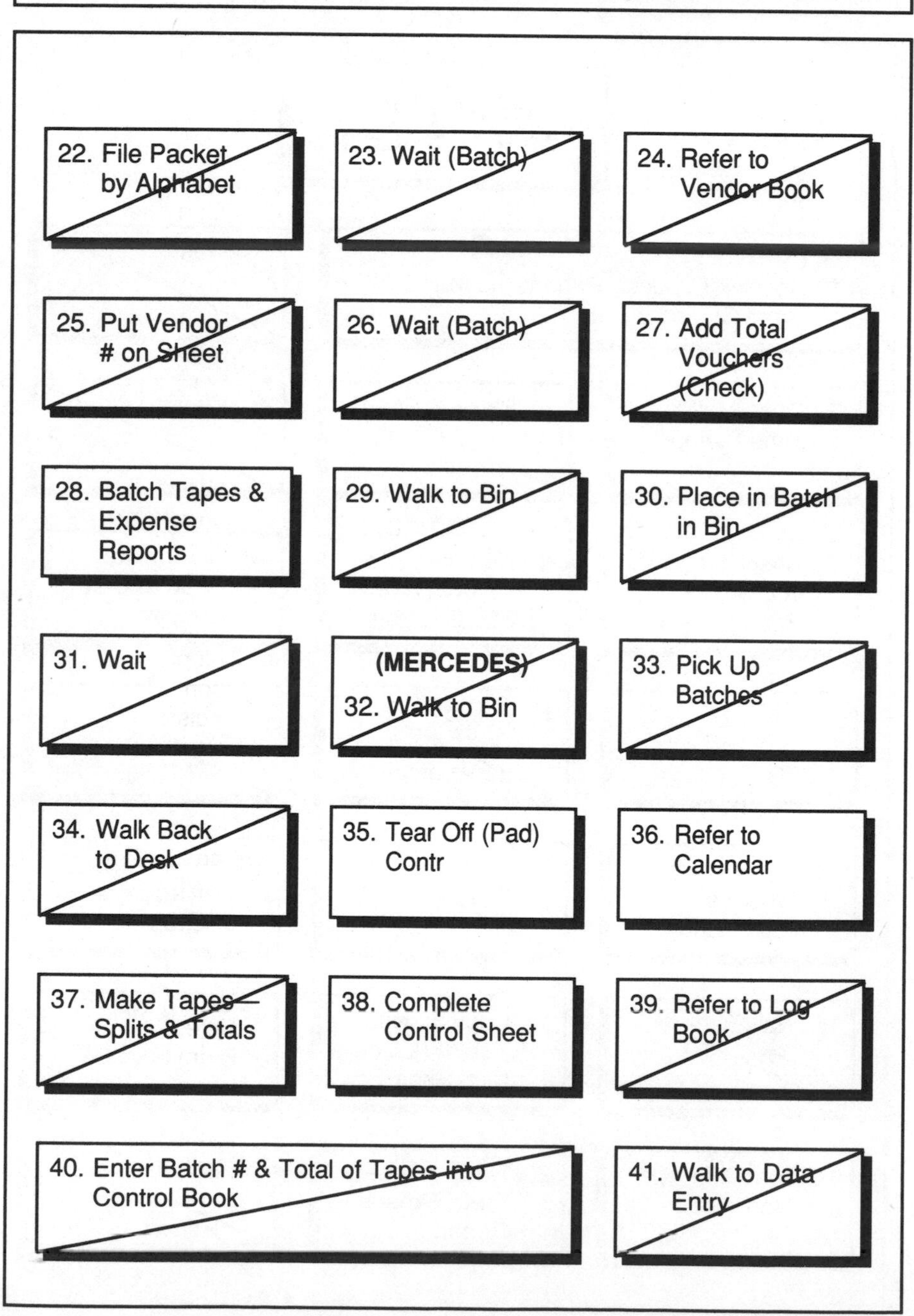
Thomas & Betts
Accounts Payable Administrative Improvement Team
22. File Packet by Alphabet
23. Wait (Batch)
24. Refer to Vendor Book
25. Put Vendor # on Sheet
26. Wait (Batch)
27. Add Total Vouchers (Check)
28. Batch Tapes & Expense Reports
29. Walk to Bin
30. Place in Batch in Bin
31. Wait
(MERCEDES)
32. Walk to Bin
33. Pick Up Batches
34. Walk Back to Desk
35. Tear Off (Pad) Contr
36. Refer to Calendar
37. Make Tapes— Splits & Totals
38. Complete Control Sheet
39. Refer to Log Book
40. Enter Batch # & Total of Tapes into Control Book
41. Walk to Data Entry

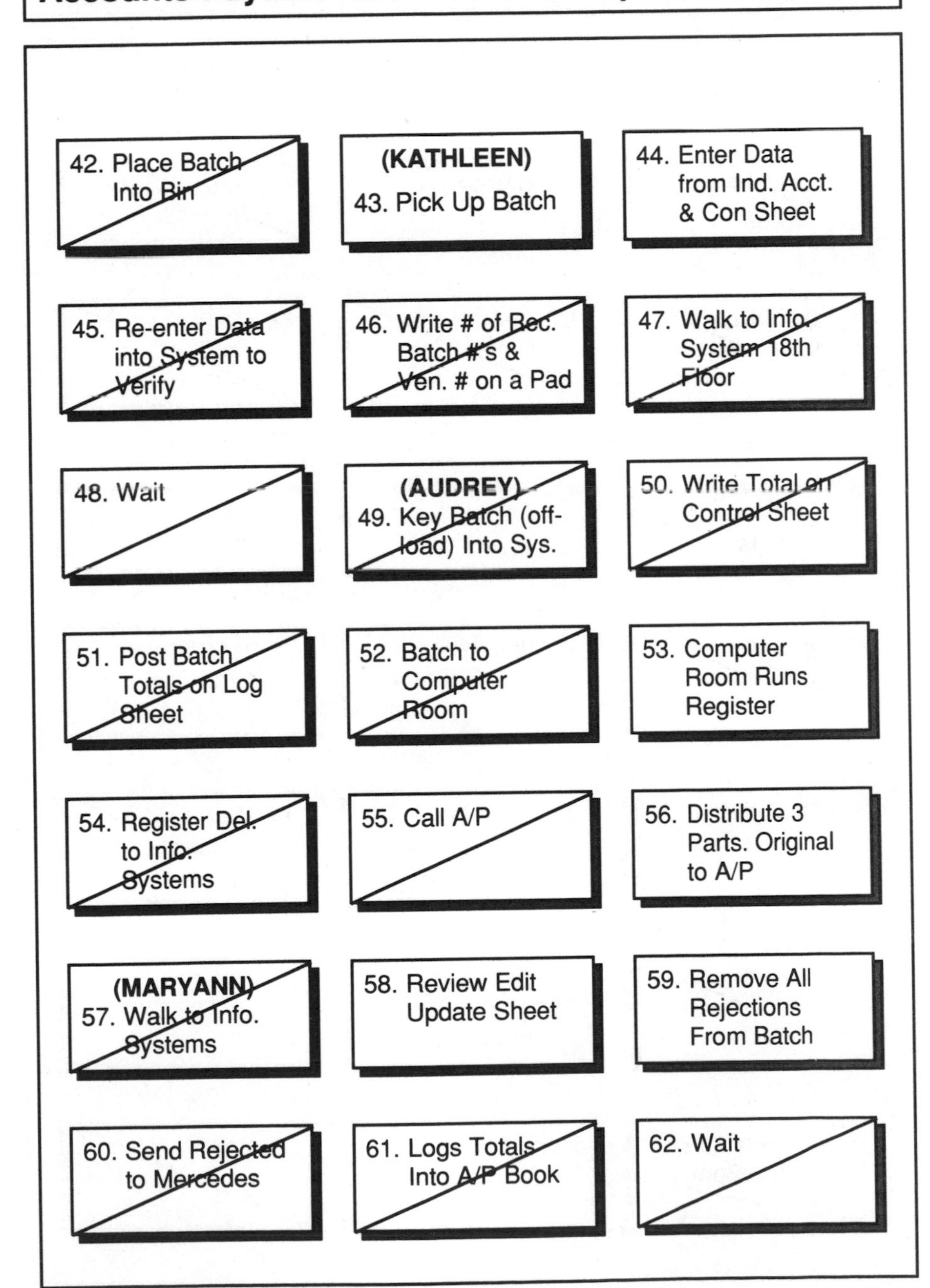

Thomas & Betts
Accounts Payable Administrative Improvement Team
42. Place Batch Into Bin
(KATHLEEN)
43. Pick Up Batch
44. Enter Data from Ind. Acct. & Con Sheet
45. Re-enter Data into System to Verify
46. Write # of Rec. Batch #'s & Ven. # on a Pad
47. Walk to Info. System 18th Floor
48. Wait
(AUDREY)
49. Key Batch (off-load) Into Sys.
50. Write Total on Control Sheet
51. Post Batch Totals on Log Sheet
52. Batch to Computer Room
53. Computer Room Runs Register
54. Register Del. to Info. Systems
55. Call A/P
56. Distribute 3 Parts. Original to A/P
(MARYANN)
57. Walk to Info. Systems
58. Review Edit Update Sheet
59. Remove All Rejections From Batch
60. Send Rejected to Mercedes
61. Logs Totals Into A/P Book
62. Wait

Thomas & Betts

Accounts Payable Administrative Improvement Team

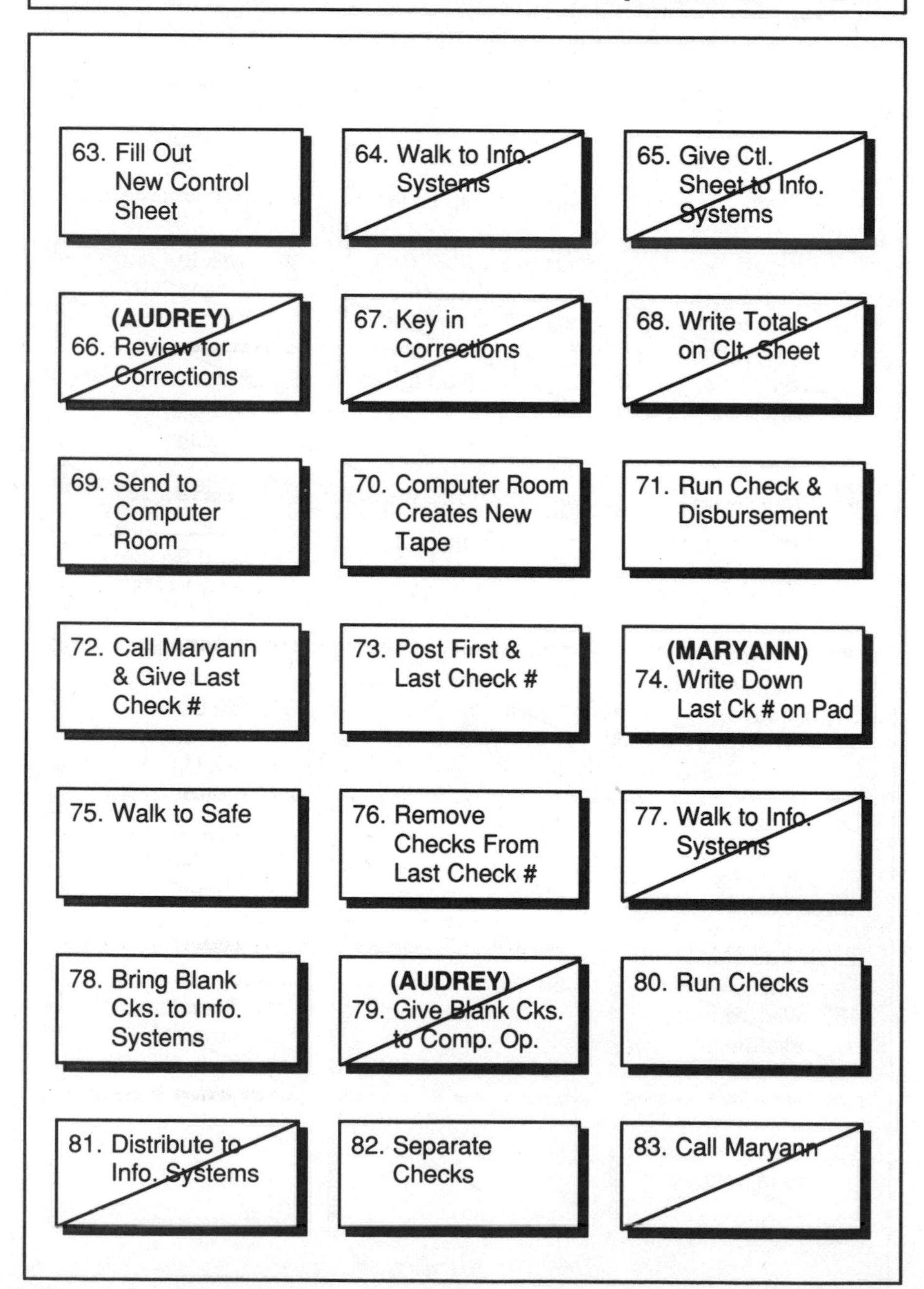

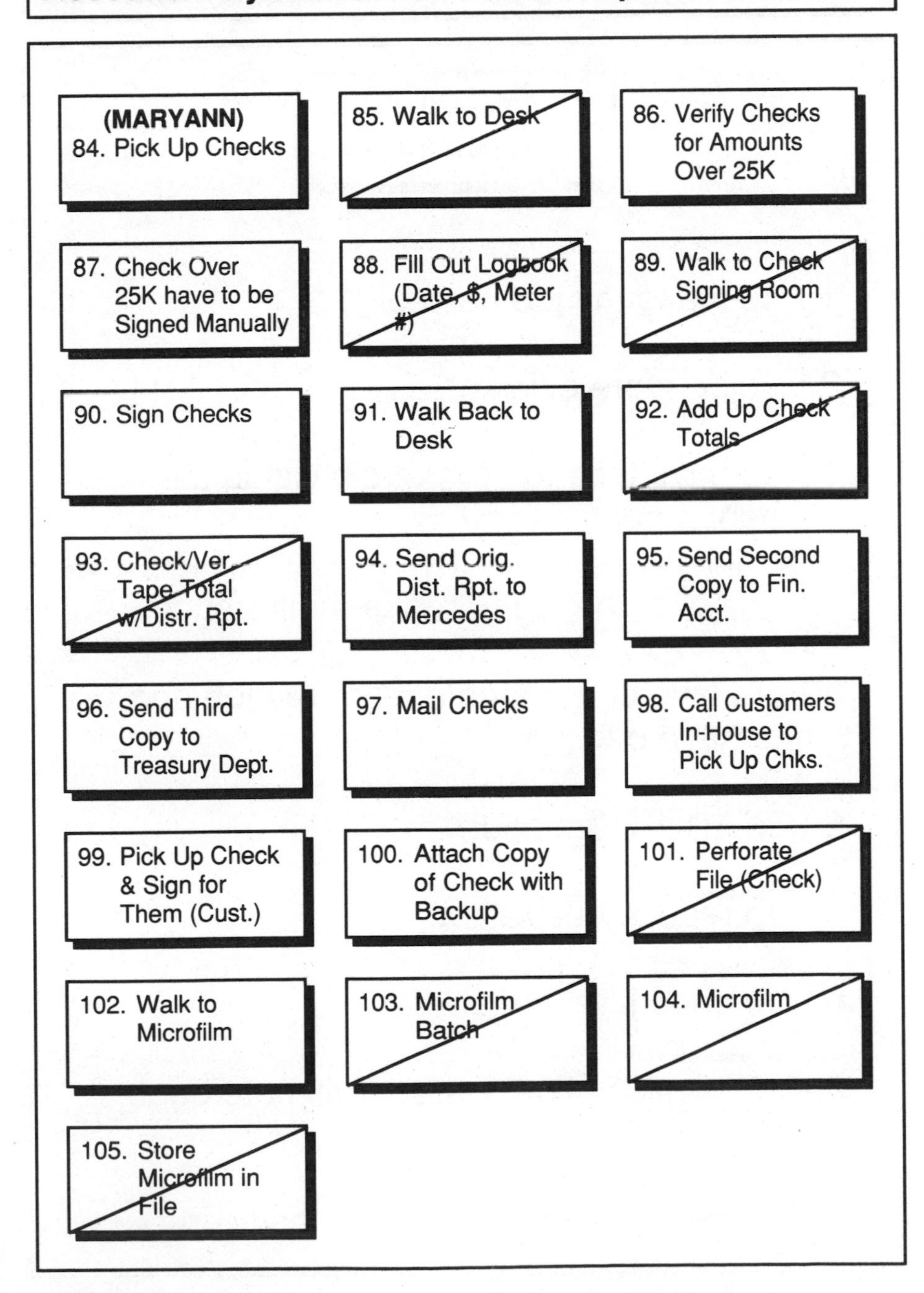
Thomas & Betts
Accounts Payable Administrative Improvement Team
(MARYANN)
84. Pick Up Checks
85. Walk to Desk
86. Verify Checks for Amounts Over 25K
87. Check Over 25K have to be Signed Manually
88. Fill Out Logbook (Date, $, Meter #)
89. Walk to Check Signing Room
90. Sign Checks
91. Walk Back to Desk
92. Add Up Check Totals
93. Check/Ver. Tape Total w/Distr. Rpt.
94. Send Orig. Dist. Rpt. to Mercedes
95. Send Second Copy to Fin. Acct.
96. Send Third Copy to Treasury Dept.
97. Mail Checks
98. Call Customers In-House to Pick Up Chks.
99. Pick Up Check & Sign for Them (Cust.)
100. Attach Copy of Check with Backup
101. Perforate File (Check)
102. Walk to Microfilm
103. Microfilm Batch
104. Microfilm
105. Store Microfilm in File

Thomas & Betts

Accounts Payable Administrative Improvement Team

Project Expectations

1.) To make my job easier.

2.) To be more efficient.

3.) To have time to cross-train within the department.

4.) To get a chance to make a difference.

5.) To have other people learn my job and to have backup.

6.) To instill team effort.

7.) To reduce frustrations.

8.) To understand paper flow.

9.) To install on-line system (ultimate goal).

Thomas & Betts

Accounts Payable Administrative Improvement Team

Work Flow Categories

Checking, Reviewing, & Logging	Walking & Moving	Calling
17	4	6
Waiting	**Filing, Batching, & Copying**	**Doing, Completing, Writing, & Filling-In**
7	6	31
	TOTAL STEPS 105	

Thomas & Betts

Accounts Payable Administrative Improvement Team

Brainstorm List of Improvement Ideas	Eliminated Steps
1. All expense reports in on the same day	4
2. Customers time-stamp own report	4
3. Send expense reports back if info. is not complete	9–13
4. Diners Club separated from Expense Report Flow	17–18
5. Change Expense Report Form	19–21
6. On-line system	19–21
7. Customers put vendors' numbers on Expense Report	24–25
8. Distribute vendors' numbers to customers	24–25
9. Locate Shirley, Mercedes, and Kathleen together	29–34, 41
10. Pad with a running Julian Calendar	35
11. Eliminate logbook	39–40
12. Access to Audrey's system	49–52, 54
13. Data entry complete steps 58 & 59	58–59
14. Send batch to Computer Room	69
15. Data entry knows last check number	72
16. Checks to Computer Room	78–79
17. Relocate check signing machine in AP	89
18. Access to Microfilm upstairs	102–103
19. Microfilm and AP located together	104–105

Thomas & Betts

Accounts Payable Administrative Improvement Team

Time, Distance Rejects

Time – How Many Days Does It Take to Process?

Sample of 7 Expense Reports	
Beginning	End
11/09	11/11
11/22	11/30
11/28	12/02
12/05	12/12
12/07	12/14
12/09	12/14
12/12	12/14

From the 7 Sample Expense Reports the Average: 2–7 Days

Distance – How Many Steps From Start to Finish?

Distance = 6765 Feet = 1-1/4 Miles for Total Steps to Process Expense Report

Thomas & Betts

Accounts Payable Administrative Improvement Team

Rejects – How Many Exp. Reports Kicks Out of a Batch?

Month	Monthly Average	Daily Range	
		Low	High
October 1988	5.11%	3.30%	11.22%
November 1988	4.60%	0.00%	14.00%
December 1988	5.53%	5.20%	18.00%
January 1989	10.70%	2.31%	23.71%

Thomas & Betts

Accounts Payable Administrative Improvement Team

Year-to-Date Results

Cycle Time:

Sample 44 Expense Reports: **40–2 or less days**
4–2 or more days

Rejects:

0%
4,000 to 5,000 per month

Value Added:

Eliminated 34 nonvalue-added steps

Exhibit 10-2. Reliability assurance program.

HUTCHINSON TECHNOLOGY, INC.

RELIABILITY ASSURANCE

PHASE I

PURPOSE:

To define the current operation in terms of products, processes, customers, and vendors involved in fulfilling the mission of the department.

REQUIREMENTS:

A. Determine what products/services are provided by your department.
B. Identify and document your process flow.
C. Estimate the unit output of your product/service.
D. Estimate the unit cost of your product/service.
E. Determine the internal and total throughput time of your process.
F. Identify the customers of your department.
G. Meet with your customers to determine the product/service specifications.
H. Estimate the outgoing quality level of your product/service.
I. Estimate the internal quality level of your process.
J. Determine what components are required by your process.
K. Identify your vendors.
L. Determine your component specifications.
M. Estimate the incoming quality levels of your components.
N. Communicate component specifications to your vendor.
O. Develop and implement performance measures.
P. Quantify and document any changes that you have made.

RELIABILITY ASSURANCE

PHASE II

PURPOSE:

To control and improve current operations.

REQUIREMENTS:

Product Definition

A. Product specifications documented and controlled.
B. Outgoing quality measurement system established.
C. Outgoing quality level goals developed.
D. Outgoing quality improvement plans developed and initiated.
E. Outgoing quality assurance procedures developed and initiated.
F. Internal quality measurement system established.
G. Internal quality goals developed.
H. Internal quality improvement plans developed and initiated.

CFM

I. CFM improvements initiated.
J. Unit cost, labor/unit, and throughput goals developed.

Component Definition

K. Component specifications documented and controlled.
L. Incoming quality measurements and feedback system developed.
M. New component prototypes approved.
N. Vendor's capability assessment complete.

Systems

O. Task breakdowns written.
P. Individual development plans initiated.
Q. Disaster recovery plan documented.
R. Safety review complete.
S. Security review complete.
T. Management control system initiated.

Flow

U. Process flow chart updated.

RELIABILITY ASSURANCE

PHASE III

PURPOSE:

To maintain improvements and reset goals.

REQUIREMENTS:

A. Operator training program complete.
B. Process capability studies complete on each operation.
C. Critical processes under statistical control.
D. Nonconforming components/material control in place.
E. Outgoing quality levels/goals met.
F. Internal quality goals met.
G. Unit cost and volume goals met.
H. CFM process flow established and operating.
I. Final CFM layout complete.
J. Inventory control system operating.
K. Management control system documented and operating.
L. Disaster Recovery Plan tested and amended.
M. All goals re-established for ongoing improvement.

Chapter 11

Eight Steps for Implementing JIT Purchasing

JIT Purchasing is the first outside-the-factory issue I will discuss. This building block, often misunderstood, is a key to the success of implementing JIT inside the factory. The eight steps for successful implementation of JIT Purchasing will be detailed, followed by a discussion of how to time the implementation of JIT Purchasing in coordination with the implementation of JIT inside the factory. The benefits of JIT Purchasing for both your company and your suppliers will be outlined. Finally, I will discuss Electronic Data Interchange (EDI), systems purchasing, and how JIT principles can improve the purchasing process.

JIT Purchasing seems to be one of the most misunderstood pieces of JIT; therefore, it is appropriate to define what it is not.

JIT Purchasing is *not*:

- A complete Just-In-Time effort
- Only for those suppliers with whom you have real clout
- A barely-in-time, frantic delivery program
- "JIT warehouses"—having the supplier stock parts instead of the customer
- Daily or hourly delivery
- Pressuring or threatening the supplier
- Only for those who have suppliers close by
- Zero inventories

Since most companies do not have their suppliers close by and many do not constitute a major portion of their suppliers' business, some managers assume JIT and JIT Purchasing won't work for them. They assume it will cost more for suppliers to deliver more often. They assume that the only way a supplier can deliver this way is for them to stock the inventory close by. Through the discussion of implementation steps for JIT Purchasing, you will see why these assumptions are inaccurate.

JIT Purchasing allows your company to have the right parts at the right time with the right quality from suppliers through a mutually beneficial relationship with suppliers. This approach supports the implementation of JIT inside a company. It is a different way of thinking about purchasing. This definition of JIT Purchasing will be expanded as I discuss the eight steps for implementation.

Step 1: Create and Educate JIT Purchasing Teams

As my colleague Ed Hay explains, purchasing professionals in a world-class organization have only three responsibilities: sourcing, pricing, and developing continuous-improvement relationships with suppliers.* Unfortunately, most purchasing professionals today are involved in the day-to-day ordering, expediting, fire-fighting, and parts-chasing activities required to keep manufacturing operations going, and there is very little time left to focus on these three important responsibilities.

In order to allow focus on sourcing, pricing, and developing supplier relationships, you should separate these functions from the day-to-day support functions. Buyers/planners can take on the ordering function with the supplier as they determine what and how many are needed and when. They can order from predetermined sources under predetermined JIT contracts or through agreements as negotiated by a purchasing specialist.

Input from many functions within the company is needed to make the right decisions about the sourcing and purchase of material. A team, led by the buyer/planner or the purchasing specialist and including product engineering, manufacturing, quality engineering, quality, material handling, and possibly cost accounting, can be formed to support each product line. The team is a process-oriented team in that it will be ongoing as long as the product line exists. The advantage of the product line focus is that overall results of JIT purchasing will be realized more quickly as an entire product line can benefit, allowing "bottom-line" results.

Another alternative for JIT Purchasing teams is a commodity group approach. This may be necessary in companies that have purchasing as a corporate function for the most part, supporting many manufacturing facilities. When corporate purchasing is separated from manufacturing, functions cannot work as closely as a team. The advantage of a commodity group approach is that it creates more opportunities to combine business

**P&JM Review*, Vol. 10, No. 2 (Feb. 1990), p. 24.

group approach is that it creates more opportunities to combine business with few suppliers and develop contracts to buy for multiple facilities. The advantage of a commodity team approach for a single entity is greater focus and expertise per commodity. However, no particular product line experiences bottom-line results quickly.

A special team that may be needed is a task team to work on supplier development issues such as supplier recognition and measurement or the development of JIT contracts and certification procedures. The education of suppliers is sometimes handled by this team as well. These are issues that require consistency among the purchasing specialists, so a team approach makes sense.

A special team will also be required to work on the elimination of waste within the purchasing process, utilizing the approach outlined for other nonmanufacturing functions in Chapter 10.

Regardless of how teams are formed, the purchasing teams will require education about Just-In-Time and the impact of JIT on purchasing. It is important that the team understand the changes taking place in manufacturing so it will be able to provide better, and possibly different, support as manufacturing progresses. These same concepts can also apply to the process of purchasing.

Education on JIT Purchasing is also required to create a clear vision of the purchasing function in the future. Then steps to get to that vision can begin. Purchasing professionals cannot be expected to negotiate successfully with suppliers until they really believe in and understand JIT themselves. By tying purchasing pilots with manufacturing pilots, the people involved can learn through experience how JIT works in manufacturing as well as purchasing.

Step 2: Develop an Implementation Plan to Ensure Systematic Success

An implementation plan should be developed by the purchasing teams to clarify the vision for purchasing in the future and set a pace for implementation. This vision includes supplier-base reduction, supplier development, certification of suppliers, raw material inventory management, and ongoing support of manufacturing, as well as improvement of the purchasing process. Measurements should be recorded regularly to report how the effort is progressing, so measurement criteria are part of the implementation plan.

Goals and measurements for supplier performance should be developed with the supplier in order for them to be a meaningful tool for improvement. If the supplier does not understand how you are measuring his performance, he will not use the information to improve. In fact, it will probably go in the trash as received!

Six Goals to Establish With Suppliers

1. *A Focus on Mutual, Consistent, and Continuous Improvement.* Mutual means improvement efforts are developed together—supplier and customer. Consistent means improvement efforts will be reasonable—not jerky—and achievable. Continuous means effort to get better will be never-ending. No matter how well the relationship is going, both parties should continuously ask, "How can we get better?"

Goal setting and measurements are one way to keep the momentum. Another is sharing information and learning. As one company has a particular success, it can invite the other to learn from that. For example, if you achieve a 75 percent reduction of setup time in one area of your facility, invite the supplier in to learn from your success.

At Xerox, a team was trained in setup reduction specifically in order to go out and teach suppliers how to do setup reduction. Your company may not be large enough to do this, but certainly any company can share learning.

Another key to mutual improvement is taking the time to think and talk about it. This requires both companies to "stop spinning" long enough to think and plan ahead—to create a vision for the future and map out steps to get there. A hit-and-miss improvement effort may provide pockets of excellence, but a carefully planned and executed improvement effort will be long lasting.

2. *100 Percent Quality.* Quality at the source—100 percent ownership by 100 percent of the people—must be the prevailing mindset in both the supplier company and the customer company. Typically, you expect 100 percent quality material from your suppliers. But do you typically give yourself the same stringent requirement in what you provide to the supplier? Do you have 100 percent quality in how you order parts, produce parts specifications, and communicate with suppliers concerning your requirements? Immediate feedback of quality information can help the supplier focus on and solve a problem quickly.

Communication—A Two-Way Street

Recently, I noticed someone in a manufacturing facility reworking parts. I asked if the supplier had been notified of the problem. This person replied, "Oh, we've been reworking this part for years." It had become a part of the process of building the product! Everyone had been in so much of a hurry to ship product that no one had time to communicate with the supplier about this problem.

3. *Continuity.* Continuity refers to regular, consistent communication between customer and supplier. As discussed in Chapter 5 on Drumbeat, regular and predictable always works better than uneven and unpredictable.

Good Communication Means Good Service

At Hewlett-Packard, I was responsible for purchasing material for a new product we were introducing utilizing the JIT concepts and techniques. I found that simply establishing regular and predictable communication with suppiers improved their service to us.

I divided my major suppliers into five groups—one for each day of the week. Every Monday, I communicated with all of my Monday suppliers, every Tuesday with the Tuesday suppliers, and so on. This communication might be a purchase order, a new forecast, general communication about existing orders, new ideas to share, future product designs, or any number of other things. But whatever it was, this regular communication forced both the supplier and me to stop spinning long enough to talk—heading off many problems.

Supplier forecasts can reduce the lead time from your suppliers significantly. A typical forecast may look something like Figure 11-1.

This forecast is usually supported by an agreement, or contract, about how much the actual quantities can vary from this forecast over time. This variability agreement needs to work in harmony with the schedule stabilization discussed in Chapter 5. Figure 11-2 depicts this principle.

Once your company agrees on a schedule stabilization policy such as that shown in Figure 11-2, the decks are cleared to begin to develop

Figure 11-1. Supplier forecast.

WEEK

PART NO.	1 DAILY					2	3	4	5-8	9-16	17-26
12345	5	5	5	5	5	25	25	25	100	200	300
13579	4	4	4	4	4	20	20	20	80	160	180
23456	40	40	40	40	40	200	200	200	800	1,600	1,600
24680	2	2	2	2	2	10	10	10	200	300	500

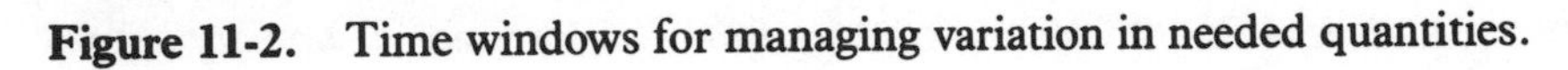

Figure 11-2. Time windows for managing variation in needed quantities.

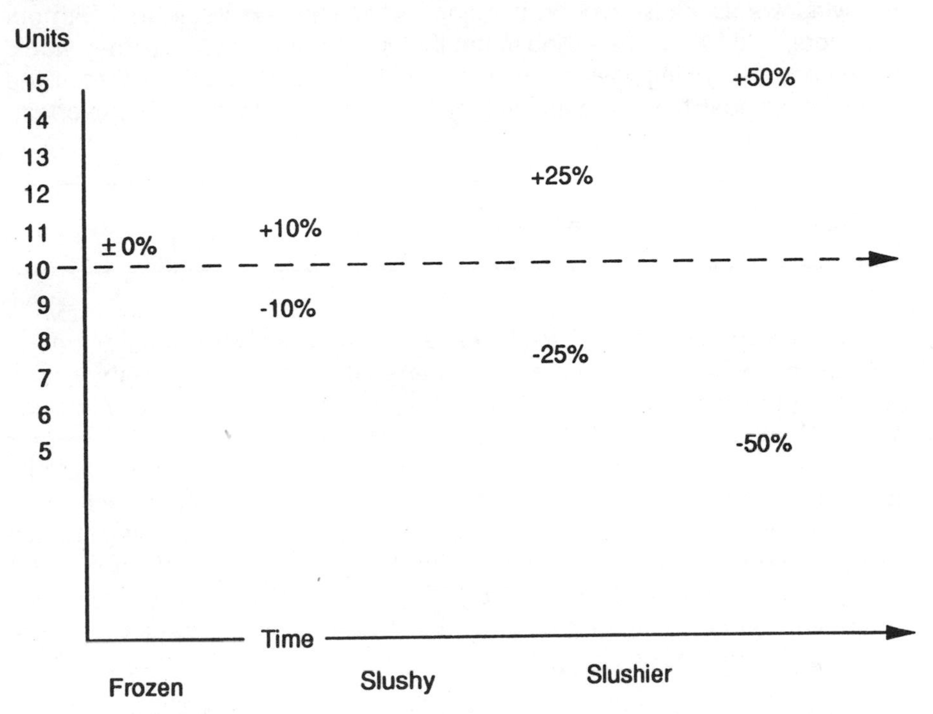

supplier forecasts. If the internal schedule stabilization policy is not created first, chances of adhering to a supplier forecast within certain boundaries are almost zero, because there is no guarantee of continuity between your build schedule and what you are forecasting to suppliers as needs for the future.

Mutual Support Builds Trust

I recall the days when personal computers were just becoming popular. The forecasts predicted that sales would zoom—it seemed as if every household would have two PCs by the holiday season that year! At the time, I was in the printed circuit board plant at HP, which provided certain printed circuit boards for the HP personal computer.

We had an agreement with our suppliers similar to that shown in Figure 11-2. Our supplier forecasts reflected the zooming sales that were expected and planned for. But sales didn't zoom as all computer companies had planned—instead, there was only a consistent upturn.

We bought and received material from our suppliers to adhere to the forecast and agreement until we could adjust our forecast downward within the windows. Because we did this, the relationship was solidified with our suppliers, and I would be willing to bet that in an upturn situation, they would have done everything possible to provide extra product above the forecasted level if at all possible—an outgrowth of the mutual trust that had developed.

Good Forecasting Will Reduce Lead Times.

If the supplier is guaranteed a certain purchase within some window, he will be willing to purchase raw materials and plan capacity before receiving the actual order. In other words, if you forecast to a supplier that you will need one hundred pieces, plus or minus 10 percent, in four weeks, you are guaranteeing a purchase of 90 to 110 parts. If the supplier's lead time for raw materials is four weeks, he will probably be willing to purchase raw materials now to prepare for your order, which will be frozen in two weeks. In fact, many companies have a rule: They do not create change orders for purchase orders. This is made possible by releasing forecasts regularly and not issuing the purchase order until the freeze period, or period of no changes.

Forecasts That Reduced Semiconductor Lead Times

During one of the semiconductor shortage periods, suppliers of those components were quoting a fifty-three week lead time. Because we at HP were providing a supplier forecast with a guaranteed minimum buy, we were getting this product in twelve weeks! The key to shorter lead times from suppliers is supplier forecasts that are real.

The benefits of continuity are clear and compelling:

- Better identification and resolution of problems
- Easier relationship building with the supplier
- Supplier lead times reduced

4. *Have the right amount of materials at the right time.* A basic premise of JIT is to learn to build the right parts at the right time—when they are needed. In order to do this, the right amount of material must be received from the supplier at the right time.

As noted earlier in this chapter, JIT Purchasing does not mean getting material just barely in time for use. Perfecting a schedule with suppliers for the delivery of parts may take years. A delivery system that delivers every part just as needed may not even make sense for all parts.

The key to success here is to establish predictable delivery with suppliers as part of the both-win relationship, then to apply the continuous improvement approach to delivery—receiving constantly smaller batches more often until the lowest possible quantity that makes sense is achieved. Once a Drumbeat has been successfully established internally, the supplier can also join the Drumbeat. This is JIT delivery.

5. *A Fair Price.* Price has typically been the number one deciding factor in choosing suppliers. The lowest bidder gets the business, regardless of many other factors, such as quality, delivery, financial stability, technical contribution, and willingness to improve and develop a both-win relationship.

Government suppliers are plagued by this issue. In order to select a supplier other than the lowest bidder, a great deal of extra paperwork plus additional signatures and approvals must be obtained. It can only be hoped that in the future this will change and a complete evaluation of suppliers will be done.

A fair price allows the supplier to receive a fair profit. As the supplier becomes more efficient, his savings can be passed on to the customer while the supplier maintains that profit. The both-win relationship ensures that this will happen. A fair price will yield total success over the long term, because total cost will be lower. For instance, using the lowest priced supplier may ultimately lead to much higher costs because of poor product quality and unpredictable deliveries. Therefore, your total cost will be driven upward as a result of part shortages and the additional inspections and rework required.

The right few suppliers will offer a fair price based on their cost plus a fair profit without threats or pressure—it's part of the partnership.

6. *Early Involvement in Design.* JIT Purchasing extends far beyond the products and parts needed today to future products and designs. After all, a partnership, or both-win relationship, is ongoing. Both companies are looking at long-term gains through this relationship. Early involvement in design is the real essence of a partnership between supplier and customer.

Suppliers can play a key role in the design of products and can provide expertise in their particular industry. No one company can develop expertise in every technology required to build their products, so all companies are dependent on the expertise of their suppliers. The key is to tap this resource early in the design cycle and to have regular design reviews with the suppliers throughout the cycle.

One important reason to choose a particular supplier is technical expertise. Early involvement of suppliers in design indicates that the supplier is willing to share technology and ideas with the customer. There may even be opportunities for joint development of new products. In some cases, leading-edge research can be done together.

In other cases, early involvement in design may indicate involvement in a change to an existing product or part. Sometimes problems are found in existing products long after they are released to production. Other times, new and improved ideas are applied to existing products. In either case, the supplier affected should be involved in the design.

Lack of Communication Can Really Cost!

This element really hit home when I was purchasing parts for a new product at HP. I received a specification from the engineer for a particular part that had very tight limits. Of course, I did not question the engineer's specification. After all, I am not an engineer and knew very little about the technical requirements for the product. So I simply looked to the selected supplier for that type of a part and placed a purchase order.

I had no other communication with the supplier. A few weeks later, when the parts were due, they still had not been received, so I called the supplier to inquire about the parts. The reply went something like this:

"Your parts will be two weeks late. We had to purchase a new tool to provide the tight specification that you require, and the tool just came in today. You asked for plus or minus one millimeter, and our standard tools can only provide a guarantee of plus or minus two millimeters."

I immediately thought of price and commented that this tight specification must have impacted the price of the parts as well. The supplier indicated that the price was forty percent higher because of this requirement! I hung up the phone and began to analyze the situation.

I decided to talk to the design engineer. During the discussion, he indicated that plus or minus two millimeters might have been okay. The tight specification was chosen arbitrarily, just to be sure.

Because I had not established a both-win relationship through regular communication, we paid more for late parts! If this relationship had been established, the supplier and I would have discussed these specifications during one of our regular conversations; the supplier would have told me that it could build plus or minus two millimeters less expensively and more quickly. I then could have queried the engineer before the parts were built.

Step 3: Pick the Right Few Suppliers for the Partnership

In the past, the relationship between customer and supplier has been adversarial. The customer has said, "You'll deliver these parts or else," while the supplier has replied, "You'll get the parts when I can work it in." At times, the relationship between the supplier and customer becomes a contest to see who can get the upper hand. Suppliers comply with customer demands only because they face the threat of loss of business if they don't. And customers are just nice enough to suppliers to make sure they are not cut off.

Developing a Both-Win Relationship With Suppliers

A both-win partnership means that the supplier and customer work in harmony. It does not mean that nothing will ever go wrong, but when it does, a solution is developed together to solve the problem. An attitude of close cooperation, proactive communication, security, and trust prevails. No secrets are kept. There is long-term commitment on both sides. The supplier and the customer develop mutual goals and enjoy mutual benefits from the relationship.

A both-win partnership is not developed overnight; it takes a lot of work and time, with determination and dedication on both sides to make this work. And this relationship cannot be developed with hundreds of suppliers; for a company to develop close working relationships over time requires working with the right few suppliers.

There are two pieces to this equation: *right* and *few*. It's important to pick the right few suppliers to work with, consolidating business whenever possible. Not only don't you have time to work with dozens or hundreds—perhaps even thousands—of suppliers to build long-term both-win relationships, but if a given supplier is getting only a small piece of your purchases in a particular area, there is no incentive for the supplier to engage in the kind of work necessary to develop a both-win relationship. Business can be consolidated with the best few suppliers, providing a better opportunity to develop a both-win relationship with one or two suppliers while giving those suppliers more business. The "right" suppliers may be different from any present ones—do not forget to analyze all possibilities as this effort begins.

How to Determine If Single Sourcing Is Right for You

Single sourcing—having only one supplier for a particular part—has long been a controversial issue. Reasons for multiple sourcing include:

"The supplier may have problems, and I need a backup just to be sure."
"Having multiple suppliers gives me an edge when negotiating price. A single supplier would surely take advantage of me."
"What if the supplier's plant burns down?" [my favorite]

A both-win relationship eliminates all of these concerns. Problems are anticipated and solved together by sharing all available information at the earliest possible time. This might include issues such as financial difficulty or the possibility of a strike in a union facility.

Valid reasons for multiple sourcing are capacity constraints and percent of total business involved. There may not be enough capacity at any one supplier to provide the volume needed, in which case it may be necessary to have more than one source. Not all companies want to be a large percent of any supplier's total business. While some companies want to control the supplier, others don't want to be such a major part of the supplier's business that they could affect the success of the company if material requirements change.

Price gouging will not happen if the relationship focuses on mutual goals and benefits. As the supplier improves and gets more of your business through consolidation, his cost will decrease. This will be passed on to you with reasonable profit as a better price.

Single sourcing represents a strong commitment and partnership to a supplier.

And how often does a plant burn down?

The Criteria for Picking the Right Suppliers

Choosing suppliers requires consideration of many factors. The supplier development task team can add criteria as appropriate as it begins to pick the right few suppliers. Cooperation is necessary between this team and the product-line teams as supplier selection criteria are developed.

Criteria to be considered in the selection of new suppliers for new products or in supplier-base reduction include:

1. *Quality*. Can the supplier provide 100 percent quality? Are quality controls in place to ensure 100 percent in the future as requirements change?
2. *Delivery*. Can the supplier always deliver when it says it will? What efforts are in place to ensure that the supplier can accommodate a Drumbeat with smaller lots more often in the future?
3. *Responsiveness*. Are the lead times short and coming down? Is the supplier flexible enough to accommodate changing requirements that are inevitable in your industry?
4. *Location*. Is the supplier as close as possible? What are the freight requirements?
5. *Size*. What is the size of the supplier? How much will your business consume of the total capacity? Can the supplier accommodate requirements for the future if business increases?
6. *Financial Stability*. Is the company financially stable? What are the risks of doing business with it?

7. *Technical Competence.* Is the supplier the kind of company that can share technology? Will it be proactive in the design of your future products? Can it accommodate future designs technically?
8. *Replaceability.* How easy will it be to find another supplier to provide the same material? This issue is particularly important when rare material is required. It could influence how diligently you are willing to work to develop a supplier.
9. *Price.* Is the supplier trustworthy to always charge a fair price, collecting a fair profit but willing to pass along savings as it improves?
10. *Improvement effort.* Does the supplier have a JIT/TQ effort in its company? Will it be able to improve at a pace that can support your own effort?

Steps to Begin to Evaluate Your Suppliers

An effective way to begin the task of selecting the right few suppliers is to make a list by commodity of current suppliers. Include the dollar amount spent with each supplier; you may be able to eliminate some suppliers quickly because of the low level of business.

The next step is for a team representing the business and technical aspects of the requirements to evaluate each supplier on the criteria selected for your company. Then begin to pick the right supplier for each commodity or speciality.

Some of the technical considerations are:

- Quality control
- Process control
- Capital equipment base
- Machine maintenance programs
- Tooling capabilities
- Technical staff
- Process documentation
- Technology specialties
- Future technology potential
- Technology history

Business considerations are:

- Process documentation
- Inventory control
- Capacity planning
- Short-term capacity availability
- Long-term capacity planning

- Material flow
- Lead times
- Order processing
- Transportation times
- Short lead time forecast program accepted
- Packaging/kitting
- Financial positioning
- Plant cleanliness
- Mutual trust
- Management support
- Safety programs
- Material procurement
- Quality awareness
- Union shop

Picking the right few suppliers is an ongoing process. As suppliers are picked, they can continue through the process of relationship development. At the same time, the selection process continues. Continuous reduction of the supplier base is necessary; this is not a one-time effort.

Reducing Suppliers at Xerox

In 1980, when it began a JIT Purchasing effort, Xerox has 5,000 suppliers. By 1985, Xerox had gone to 270 suppliers; the plan is to be under 200, and the company thinks that 150 is the right number of suppliers.

Reduction of the number of suppliers is a little like inventory reduction—it can't happen all at once, but there needs to be a continuous effort to reduce the number.

Step 4: Select Pilots to Learn the Process

The selection of the right few suppliers can occur in parallel with beginning purchasing pilots. Pilot suppliers/parts should:

- Be one of the chosen few
- Support manufacturing pilots
- Have the potential to develop skills within the purchasing organization that can be transferred as other suppliers become pilots

JIT Purchasing is a massive effort. Pilots are necessary in order to support the sheer volume of work required as well as to learn as you go. Pilot suppliers/parts that are selected with a product-line focus will be more visible and can be managed by the product-line teams. In other words, selecting several parts from one product bill of material to work on rather than having a scattered selection process that does not focus on a particular product may get more people involved in JIT purchasing.

Purchasing pilots and manufacturing pilots, when aligned, can support one another. As manufacturing has new requirements, the supplier chosen by purchasing will be ready to support those needs.

A Purchasing and Manufacturing Partnership

Recently, a manufacturing team at Valleylab encountered a problem with a certain carton. It did not close well and the product frequently fell out, causing quality problems. The manufacturing team created a couple of prototypes for a new carton with cardboard, scissors, and tape.

Since a JIT Purchasing team had selected the carton supplier as a pilot supplier for JIT purchasing, the manufacturing team was able to invite the supplier in and explain its prototype. The supplier incorporated the design suggestions; documentation changes were expedited since engineering was on the team. Today, the new carton is working well and saving the company money. The manufacturing team was able to demonstrate its influence far beyond what had ever been possible before.

Step 5: Educate the Supplier to Develop a Common Vision

Selected pilot suppliers should be educated in JIT principles so that they understand your improvement effort and what requirements that will place on them. They may also be required to implement JIT principles in order to meet those requirements. There are many ways to accomplish this; you will have to decide which methods work best for you. The important thing to remember is that you can't expect suppliers to behave differently until they understand the vision for the future and the benefits for both you and them. Education and training for suppliers is not unlike that required in your company; it should include the same concepts and principles as emphasized within your own company.

One method for educating suppliers is to have them visit your facility. This is often called a "supplier day." The agenda is usually to tell the suppliers what your company is doing and what the vision is for the future. It's important that they get a clear picture of the requirements that you have for suppliers and how you plan to proceed in implementing them.

There are some very significant advantages in having the top management of a supplier company visit your facility. First, its top people hear your message from your top people. Second, managers can see firsthand how their product is used in your product. This may be the first step in sharing technology!

A Supplier Visit Leads to Improvement

When Valleylab first selected pilot suppliers for JIT Purchasing, a supplier day was held. As one supplier was touring the generator area, he noticed someone chiseling a bolt out of the part his company supplied. This was a very noisy and time-consuming process that took place for every generator produced.

The supplier asked why this was being done, and the operator replied that "we have always done it this way." The bolt came epoxied in, but the piece did not need the bolt. The supplier replied that it would be easy not to include the bolt.

The result of this visit was that the supplier was able to supply the part without the bolt—at a lower cost—and save time and effort on the line.

In addition to having suppliers visit you, there is benefit in your visiting the supplier. Communication with the supplier's key people about your JIT/TQ effort will probably take place at their site because of travel logistics. This will give you an opportunity to talk to many of the people in the supplier company. Of course, regular visits to suppliers are also necessary for you to review their operation. In addition, some companies have also organized educational events for their suppliers. Others have trained internal people to go out to suppliers to educate them on improvement efforts.

Step 6: Certify the Supplier to Avoid Inspection

Certification requires 100 percent quality from the supplier. As a supplier or part is certified, there is no longer a need to inspect the parts as they are received from the supplier, for you know that they are good. Many companies require certain documentation with the parts to show that they have been confirmed as good through the supplier's own quality procedures.

Sample inspection is sometimes utilized as suppliers are certified. For instance, every third shipment is inspected for a period of time; if no problems are found, every tenth shipment may be inspected. There is a clear plan in place if a problem does occur. Response to and resolution of problems is fast and complete.

Certification requires that all documentation for the material be updated, correct, and complete. This includes drawings, specifications, inspection procedures, and any other critical elements.

Efforts aimed at 100 percent quality can begin early in any JIT effort. The first step in any certification effort is to develop a certification policy, which defines exactly what certification means as well as the procedure to be followed to certify a supplier or part. This policy will differ depending on the industry; again, there is no cookbook.

Steps to begin a certification process usually include:

1. Determining key measurements and beginning to evaluate regularly the supplier against those measurements; this process develops a baseline.
2. Visiting the supplier to evaluate fully the business with predetermined criteria that are applicable for your business.
3. Developing a mutual action plan to certify the supplier.
4. Monitoring progress until the supplier has reached the required level of performance, and then certifying it.
5. Continuing to monitor according to the supplier's success.

Some companies begin a certification effort by certifying parts individually. However, the goal is to certify suppliers and eliminate incoming inspection for regular production parts in your facility. Incoming inspection should be only for new products and parts until they can be certified. This accomplishment opens the opportunities for material to travel from receipt to the production line for use.

In picking pilot suppliers, or parts, for certification, long-term success should be considered. This means picking a variety of certification candidates. Three categories come to mind:

- Quick and easy
- Difficult, important, but time-consuming
- Support of product-line pilots

Possibly, some combination of these three categories is appropriate. If you just work on the quick and easy certifications, next year the well may be dry for new candidates.

Certification is a key part of JIT Purchasing. As mentioned earlier, this effort may require a special task team to develop the policy and drive the process. However, the product line teams (or commodity teams) will be involved in the effort.

Supplier Quality Can't Be Inspected In

As I mentioned earlier, Xerox was down to 270 suppliers by 1985. By this time, the company had reduced incoming inspection to only 5 percent of material, and today, only new suppliers and/or new commodities undergo incoming inspection. By working with suppliers, Xerox found in 1985 that twice as many quality problems were occurring in material that passed incoming inspection as in material controlled properly by the supplier before shipment.

Step 7: Attain Delivery Goals to Support the Drumbeat

Delivery of material from the supplier includes the aspects of timing, quantities, packaging, and freight methods.

How to Turn Predictable Delivery Into Just-In-Time Delivery

Predictable delivery is the first step—the material arrives when expected in the quantity promised. At this point, this may still be a large quantity covering several weeks' requirements, but it arrives as promised. Once predictable delivery is achieved, a plan should be developed with the supplier to begin to increase the number of deliveries and decrease the quantity in each one. This can continue as long as it makes sense.

For more expensive and large material, more deliveries make lots of sense, but for very inexpensive and small material, fewer deliveries may be just fine. The cost/benefit should be weighed at each step, always driving toward delivery just as you need the material.

As manufacturing establishes a Drumbeat, deliveries from suppliers have an opportunity also to be regular and linear. This is your key to success.

The vision is that material arrives at the dock exactly when expected and goes directly to manufacturing. Manufacturing pulls the material as needed and utilizes the material in the product. Figure 11-3 depicts this procedure.

How Packaging Plays an Important Role in Timing and Quality of Deliveries

Material packaging is one very important factor that is often forgotten or ignored. The way material is packaged can greatly affect your success in achieving the vision.

The goal is to touch the material only one time—when the material is used in the product that you are building. This goal has many implications about type of packaging as well as quantities and protection required. Some considerations follow.

Figure 11-3. Direct delivery to the line.

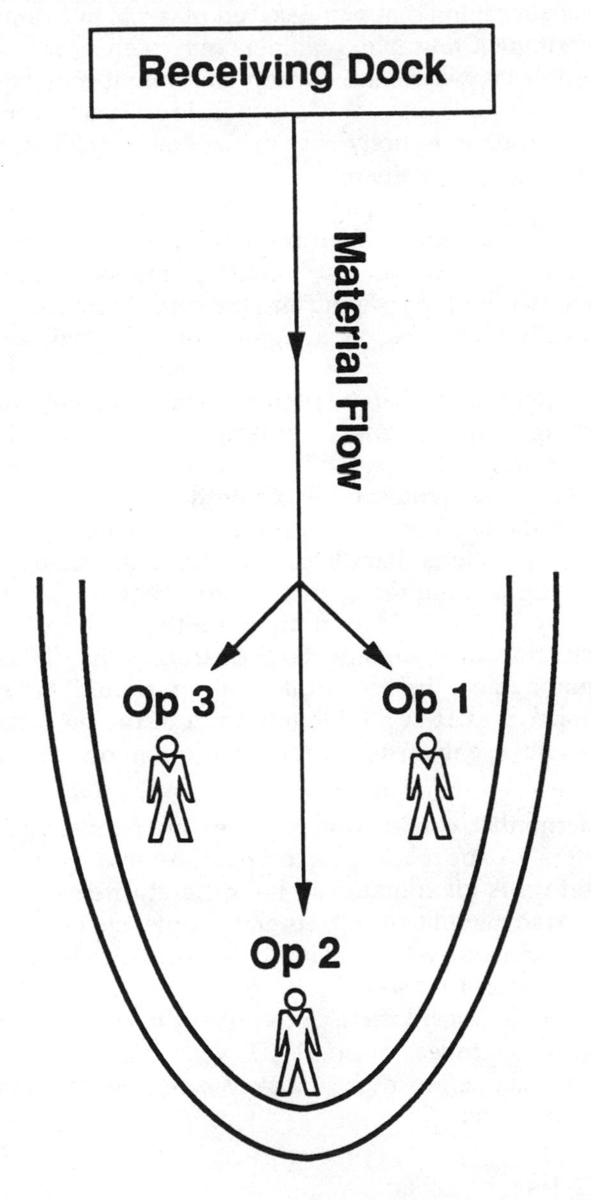

- Minimize dunnage disposal. Try to reuse packaging, or use the same packaging that you received material in to ship the product.
- Determine if reusable containers make sense. In a JIT environment, reusable containers are appropriate only if they serve as a demand signal—if the container is for specific material in specific quantities. Otherwise, an entirely new management challenge develops—that of managing containers.
- Use standard packaging whenever possible. If standard packaging doesn't make sense, then try to standardize everything else—standard sizes, closures, and handling—to reduce wasteful operations associated with packaging and moving material.
- Consider packaging as a significant cost item, and minimize the cost.
- Pay attention to safety requirements. Prevent injury to persons packing, unpacking, transporting, storing, and disposing of containers and materials.
- Protect parts. Protection is required to prevent damage that would impair the function or appearance as specified on the drawing.
- Facilitate material handling. Quantities should be usable by manufacturing without recounting or repackaging.
- Identify material. Identification should be easy to read and complete. If your company requires traceability to meet government requirements, the trace information should be supplied by the supplier as part of the identification of the part. Bar coding can be very effective in ensuring correct information and ease of collection.

One term that continually pops up in manufacturing is "kitting." Kitting refers to the packaging of parts necessary to build a unit in manufacturing. A kit contains all the different items needed. As flow is implemented in manufacturing, several people will be involved in building the unit. Kits must then be "unkitted" in order to provide the correct parts for each operation or person.

In most industries, kitting does not support JIT; the right quantities, or packaging multiples, support JIT very nicely. Only in a build-to-customer-order business might it make sense to kit parts ordered uniquely for a particular order.

Addressing Freight Costs

In most companies, freight represents a significant part of cost. Considerations should be made for freight, whether to get material to your plant to build product or to ship your product to your customer. Various arrangements are usually made concerning who pays freight, but in the long run, you pay either in higher material costs, a higher price for your customer, or lower margins for you because your customer will pay no more.

Careful management of freight can produce significant payback. For instance, trucks should be filled whenever possible—both ways. Trucks that deliver to you should leave with product being shipped by you.

Regular Parts Pickup Equals Regular Deliveries and Lowers Cost

At HP, we were able to establish several "milk run routes." After all, Boise, Idaho, isn't on the way to anywhere! No natural routes existed through Boise, but we were able to improve our freight management. These milk routes were established in areas in which several suppliers existed. Each supplier knew exactly when the truck would be by each week to pick up material. The truck would make the route once a week and deliver the material to Boise. Trucks were filled, and the freight companies loved it! Rates were lowered as a result of guaranteed regular business and full trucks.

Freight company top management personnel were included during supplier days. They too needed to understand our vision. It was amazing how much cooperation we received from the freight companies once they understood that this new plan of ours could be of great benefit to them as well.

I'll never forget one day that I stayed in the office during lunch. I was in the Materials Department at the time. The phone rang; it was a truck driver calling from an outside pay phone (I could hear the wind howling in the distance). The driver had stopped to call because he was going to miss his four-hour delivery window by a couple of hours. There was five feet of snow, and he had been delayed.

Now, this was real commitment. The top management of this freight company had done an excellent job of passing down just how important it was to have ownership for servicing the customer. The drivers clearly understood that we were depending on that material within the four-hour window or we would have to shut down the production line.

Step 8: Continue and Expand

Implementation of JIT Purchasing is a continuous process. Suppliers will be at different stages of implementation, and even after the basics are implemented, continuous improvement is necessary. Expansion of JIT Purchasing should occur at the fastest possible pace to ensure success with JIT throughout the company. Material costs represent as much as 60 percent to 70 percent of the total product cost in most companies today. No company can afford to ignore this most important aspect of JIT.

The Right Timing for JIT Purchasing

A common concern among many companies implementing JIT is timing—when to do what. What comes first—JIT in manufacturing or JIT Purchasing? The question is similar to the "chicken or the egg" question.

One of the most frequent comments manufacturing managers make about Just-In-Time is, "If I only had *good* parts on time, this JIT stuff would be simple." They believe the key to JIT is an implementation plan that looks like Figure 11-4a. On the other hand, when I talk to people in the purchasing department about implementing Just-In-Time Purchasing to support a JIT manufacturing effort, they say, "How can we possibly implement JIT in purchasing before those manufacturing people figure out what they are going to build and when? We spend almost all of our time expediting or pushing out orders with suppliers." Purchasing departments envision JIT implementation as something like Figure 11-4b.

The question, of course, is: Who is right? The answer is, both parties. Parallel efforts can begin for success with JIT (see Figure 11-5) but should be managed very carefully.

Parallel and continuous efforts in purchasing and manufacturing can be very effective. In fact, they are required to get maximum benefit. In Figure 11-6, the eight steps for implementing JIT Purchasing are highlighted. Arrows indicate critical communication and coordination points with the JIT implementation in manufacturing.

How Your Company and Your Supplier Can Reap Mutual Benefits

In the spirit of a both-win partnership, benefits should be achieved by your company as well as your suppliers. Understanding how JIT Purchasing can benefit both parties is essential for implementation. Suppliers will need to understand clearly what is in it for them. Benefits typically shared include:

- Greater flexibility
- Less space
- Lower inventories
- Regular communication/releases
- Higher quality/quick feedback
- Shared technology
- Easier sourcing for new products
- Improved manufacturability
- Shorter tooling/lead times
- Predictable delivery
- Competitive pricing

Figure 11-4. Implementation of JIT as seen by manufacturing and purchasing.

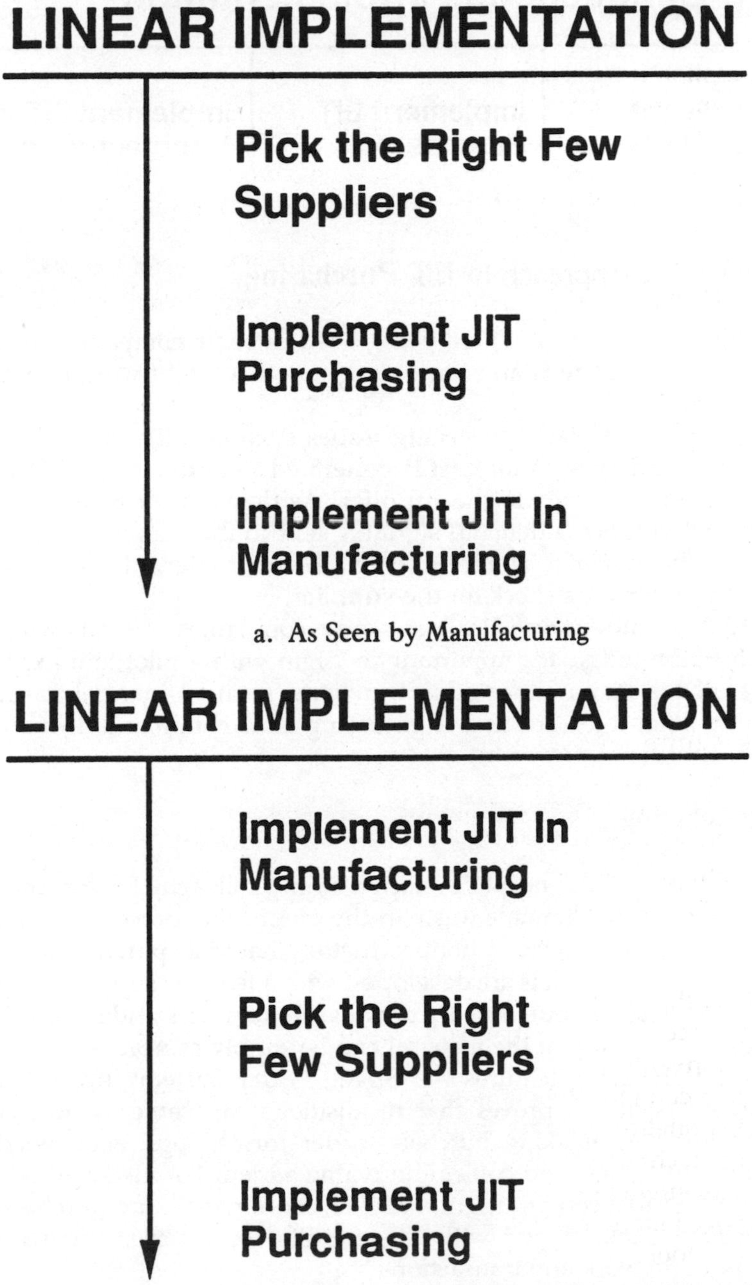

LINEAR IMPLEMENTATION

Pick the Right Few Suppliers

Implement JIT Purchasing

Implement JIT In Manufacturing

a. As Seen by Manufacturing

LINEAR IMPLEMENTATION

Implement JIT In Manufacturing

Pick the Right Few Suppliers

Implement JIT Purchasing

b. As Seen by Purchasing

Figure 11-5. Parallel implementation of JIT.

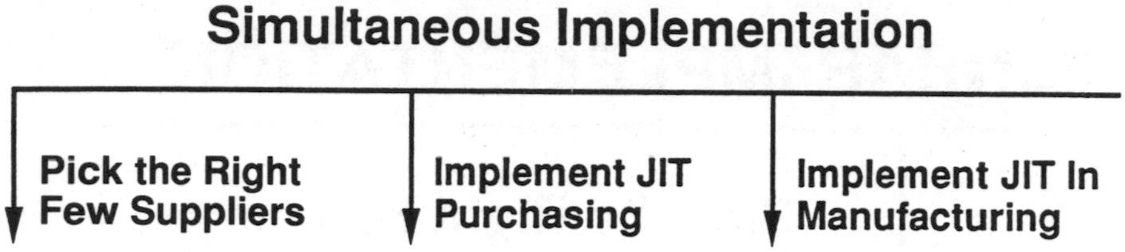

A Leading-Edge Approach to JIT Purchasing

Electronic data interfaces (EDI) are being used in some companies in lieu of purchase orders. Here is an example of how this might work between two companies.

A JIT contract is in place, governing issues such as adherence to forecasts, delivery, and price. A forecast is generated from the internal Drumbeat and sent electronically to the supplier. As demand for material falls into the freeze window, a demand signal is sent to the supplier electronically. When the material is received by the customer, the computer automatically generates a check for the supplier.

This may sound unrealistic, but many companies are following a similar procedure today. It's important to begin with a pilot and expand from there. EDI is also used for other functions, such as financial transactions, quality data, forecasts, and engineering information and documentation.

Systems Purchasing

Systems Purchasing is a concept developed by my colleague Ernest Anderson many years ago to eliminate waste in the purchasing process. Through the years, it has been applied to nonproduction materials purchasing with significant success. Contracts are developed with a few key suppliers; since most of this material is purchased from distributors as standard off-the-shelf items, the demand for the material can be simply as needed.

Systems Purchasing features the Speed Order System® (SOS),* the ultimate in simplicity. It proves that requisitions can serve as purchase orders and receivers, making purchase order forms, logs, acknowledgments, open-order files, and copies, nonvalue added. For instance, when pencils get down to a certain inventory level, more pencils are purchased. This is really a demand pull with the supplier of pencils, eliminating unnecessary paperwork and transactions.

*Speed Order System is a registered trademark of Ernest Anderson.

Figure 11-6. Parallel efforts in purchasing and manufacturing.

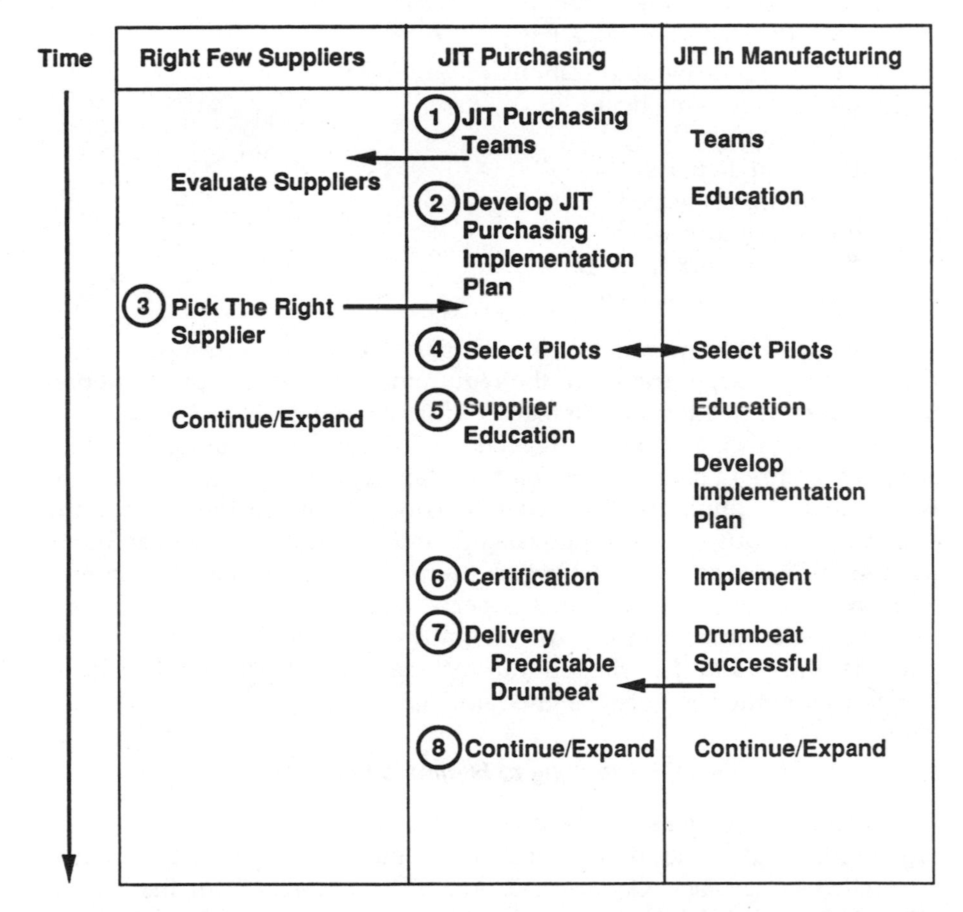

Systems Purchasing does not get into the both-win relationship as JIT Purchasing does, but the idea of eliminating waste is the same. The concepts learned through the implementation of Systems Purchasing for nonproduction material are applicable for JIT Purchasing today. Therefore, opportunities to eliminate waste in the process of purchasing both production and nonproduction material abound.

How to Benefit From JIT Principles in the Purchasing Process

The typical purchasing operation is rife with waste; since purchased materials account for 60 percent to 70 percent of the cost of most manufactured goods, eliminating waste in the purchasing process can have a huge payback. Typical wastes found in most purchasing processes include:

- Paperwork
- Duplicate records of transactions
- Change orders
- Acknowledgments
- Multiple communication links
- Material return authorizations
- Central receiving docks
- Counting/Weighing
- Receiving Reports
- Incoming inspection
- Stockroom put away
- Stockroom picking of parts
- Detrashing of packaging

The process required to provide parts to manufacturing and the facility—from the generation of the requirement to the receipt of the part by the user—is ripe for the elimination of waste in most companies.

For instance, a standard procedure for receiving material from a supplier requires a computer transaction that generates paperwork. Paperwork must be duplicated so that one copy can be included with the material, one copy can go to purchasing, and one copy can go to accounts payable. When accounts payable gets the paperwork, it must be matched with the purchase order and other paperwork. All copies of everything are filed! We have actually verified that in many companies, the same piece of paper is copied and then filed in four different places. This same transaction is also on the computer. Sound familiar?

Reorganizing to Eliminate Waste

The Valleylab materials organization began to analyze the way it was organized; it was a traditional organization of buyers, material planners, and production schedulers. On average, it took two weeks from the time a requirement was identified by the material planner to the time that the purchase order was placed. And that was when everything worked! Occasionally, paperwork would get misplaced and the purchase order would not be generated, creating a part shortage.

The paperwork was probably created to maintain control at some point in the past. These old-fashioned communication links can be replaced by giving one person total ownership of a complete process, eliminating confusion, wasted time, and mistakes.

Valleylab solved this problem by reorganizing into buyer/planners. Now, each buyer/planner plans and purchases material for a particular product line. The communication link has been eliminated.

By aligning purchasing with manufacturing, a team can be formed to provide full support for a product line. A buyer/planner has the opportunity to really understand the product he or she supports. Eventually, the production scheduler position can also be included in the buyer/planner role. Schedules are created to a Drumbeat, or a daily schedule, and the buyer/planner is now connected to what is happening in manufacturing on his or her product line. There is no need for detailed scheduling of operations, since flow has been implemented. The people in manufacturing monitor their own balance and flow, informing the buyer/planner of any exceptions.

Because this is the way of the future, it is important that materials professionals in your organization begin to plan for this reorganization and that they develop the skills necessary to do all three functions. The result is an enhancement of their job; nobody loses with the new system.

As you begin a transition to this new organization, it is effective to pair up buyers and planners to train each other. This team of two has the shared responsibility of the buyer/planner. As each learns the entire job function, each begins to take on the entire process with his or her own product line. This procedure makes for an effective and less disruptive transition. Of course, it will also be necessary to provide more formal training classes, covering the different functions of planning, purchasing, and production scheduling. Some of these classes can be presented by different members of the group.

In some organizations, there are a few senior people—the purchasing specialists—who work on the development of new suppliers and contracts. This role requires strong negotiating skills. The long-term goal should be to develop every buyer/planner so he or she can develop suppliers and negotiate contracts. As each buyer/planner now has fewer suppliers and total ownership of the process, the opportunity is much greater to develop these skills, and the materials professional can become a member of the manufacturing team, rather than a support person who supports many product lines and does not feel a part of any.

Using a Value-Added Analysis for Identifying Waste in the Purchasing Process

A value-added analysis of the purchasing process will quickly identify those wasteful activities that can be eliminated. The idea here is to create a flow of information that produces the least amount of waste.

JIT Purchasing can benefit your company. As manufacturing makes progress with JIT, an effort in purchasing will be essential for continued success. To remain competitive, purchasing will be required to become supplier managers, focusing on mutual goals and improvements.

Chapter 12

A Checklist for Materials Management, Planning, and Scheduling Systems to Support JIT

In order for a Just-In-Time implementation to be successful, a number of operations systems that gather and process data necessary to run the business need to be changed, either manually or by computers. Too often a company's policy and procedural decisions are controlled by these operations systems, and the data processing department has the only key for changing the way business is managed.

You can best influence operations systems in your company by first proving the benefit of the principles of JIT, even if you must do so outside of the system in a manual mode. Then you can do a clear study to demonstrate that the present system will not support these principles and will, in fact, hinder your progress with JIT. This is a slow, painstaking process. The more you understand about operations systems, the more you will be able to influence those tools that have such a massive effect on your everyday life.

Systems should never dictate how the business is managed; people must always do that, at every level of the organization. Operations systems need to be viewed as a tool to assist in running the business the way it should be run, with data processing providing support to those who use the system and make business decisions. The problem with most operations systems today is that they have too much of the wrong stuff and not enough of the new requirements—they have almost unending ability to handle detailed data and batches and very little ability to handle flow manufacturing.

Before looking at the global attributes necessary for operations systems in a successful JIT environment—specifically, inventory, bill of material, planning, production control, and financial—I talk about how to time the analysis of and changeover from the present operations systems in a company implementing JIT. Then I present a more detailed discussion of specific requirements for each of these systems, followed by some tips on implementation.

Define the Tasks and Then the Systems to Support the Business

Systems are tools that support the way your business is managed, and tools cannot be defined until the task at hand is well defined. The vision of how the company will be managed must be determined before the systems' needs can be correctly defined. Forming this vision requires time and experience. Therefore, systems changes should begin only at the point at which the present systems impede progress with the implementation of JIT concepts and techniques.

However, a company with no formal systems in place may feel the need to define systems that support JIT before experience has been gained with JIT and a vision of how the company will be managed in the future formed. If a company decides that it must have a system just to organize data to run the business and also decides to implement JIT, it will be forced to consider future needs without knowing exactly what the future will bring.

Those who can't wait for JIT implementation can get help from companies in a similar industry that have experience with JIT to be certain that basic systems requirements will be met for the future. For those companies that have an adequate system in place, systems considerations should wait. Information systems can be thought of in the same way one thinks of automation. (See discussion on automation in Chapter 6.) Automation of data handling can increase the quality, consistency, and speed at which data are processed. However, automation also limits flexibility and institutionalizes procedures. Just as you would when considering any other use of automation, eliminate any wasteful processes before automating. As the processes are simplified in manufacturing, fewer data may be required.

Systems implementations are very time-consuming, a lot of hard work, and usually very expensive. Therefore, it could be wasteful to implement a system to satisfy short-term needs only. Some companies find that even before the system is fully implemented, many of the system's features are obsolete.

Operations Systems Priorities and Timing

The ideal approach for operations systems is:

1. Keep existing systems.
2. Understand JIT impact on running the business through pilots.
3. Develop a vision for the future.
4. Implement systems with lasting value in a JIT environment.

Look Before You Leap

When I first joined HP, I was a programmer analyst in the Management Information Systems department, with responsibilities for manufacturing systems. The company was just beginning to look at replacing all of the old operations systems when I arrived. We in MIS investigated several systems available within HP and chose a good standard system that would meet our needs. At that point, we had no experience with JIT, nor had we ever heard this term before. However, the culture at HP allows for experimentation and trying new things.

MIS was close to completing the new system implementation, including inventory, bill of material, order processing, planning, and production control, when someone from manufacturing told me that because the new system kept track of everything, it required him to do a transaction before moving parts; furthermore, the parts were always moved into stores in order to be transacted and verified. This gave manufacturing and other departments the ability to know exactly where everything was at all times, on-line from any terminal in the facility. Accountants were elated with the control this represented. They were sure that we would never lose parts now.

But as far as the people in manufacturing were concerned, it made no sense to put subassemblies into stores, only to have them pulled back out again by the person at the very next workstation, less than twenty feet away. The person from manufacturing told me he planned to have each person just hand the parts to the next person, and I could do whatever I wanted to make the system think that he had gone through the proper steps to satisfy accounting and management.

In order to support what made sense in manufacturing and still abide by the rules of our new system, my first major systems project was to make changes so internal transactions were created to "fake out" the system. Clearly, our system was far too complicated for our simplified process.

Three Global Systems Requirements for JIT Success

There are three global requirements for business systems used in a JIT environment:

1. Simplicity
2. Flexibility
3. Responsiveness

Simplifying Systems to Avoid Waste

One of the watchwords of JIT is simplicity; this concept applies to systems as well as to actual manufacturing-floor activities. Systems in a JIT environment should be straightforward and easy to use. They should focus only on the necessary data for collection and processing. As processes are simplified throughout the organization, the systems to support those processes should be made simpler as well.

Unnecessary transactions are waste; if a system requires wasteful activities to take place in the organization, that, too, is waste. The ultimate sophistication is simplicity; simple systems enhance the ability of employees to take ownership for establishing their particular process and following their process in a disciplined fashion.

Complexity in the form of controls and data and transactions is necessary in an environment lacking in employee involvement and ownership. This complexity is generally seen as a way to check up on the people—to ensure that they are doing their jobs. The problem with this approach is that as long as this complexity prevails, people have very little flexibility to make changes or incentive to fight that inflexibility. The environment is so restrictive that to get involved and get that ownership is nearly impossible. As complexity decreases, it is necessary for people to achieve ownership for the process; they become their own check and balance on a day-to-day basis.

Simplifying Fosters Increased Ownership

In one assembly line, the employees had checks and balances built in with the complex system as parts were moved; if an employee got the count or identification wrong, there were two more checkpoints at which to identify the mistake. However, when employees took on the responsibility of handing the parts to the next person on the line, the counts and identification had to be right, because there was no stores person counting and checking behind the employee and again before the next person received the parts back from stores. This gave the employees increased accountability for the number of parts. With this accountability came ownership, and the counts were more accurate than ever before.

Decreasing complexity in operations systems should be driven by the increase of ownership within the environment to be successful.

Supporting Changes With Flexibility

Flexibility is another watchword of JIT. Continuous improvement requires continuous change. Change becomes a way of life, and the operations systems are tools to help support that way of life.

Flexibility for operations systems can be interpreted several ways. Not only does the system need to be flexible to handle changes in the way you run the business; it must also accommodate changes in the way data are input, retrieved, and manipulated.

Implementing such changes in the way the system supports the business cannot take weeks and months; the system is required to incorporate changed procedures immediately.

Easy retrieval of information is key to running the business and making good business decisions. As you change the basis for some decisions, different data will be required. Again, you cannot afford to wait while a system is reprogrammed to accommodate your need. Data retrieval and reporting will need to be on-line and under your control as a user. Easy-to-use programs that anyone can learn are a necessity in a JIT company.

A popular way to manipulate data easily or simulate situations and results is to send data from the main computer to a personal computer. There the data can be manipulated and either sent back to the main computer, deleted, or kept at the PC level only. This capability should be considered with any business system intended to support JIT.

Managing the Transition

Another company had an interesting experience during its change process. Management at this site had decided to implement flow into manufacturing, eliminating the transactions back and forth to stores. During the time it took to change to a more flexible system, five clerks were employed to input all of the data required by the traditional system to make it "think" that all of the "checks and balances" were still in place. After a more flexible system was implemented, these five people were shifted to much more rewarding jobs.

In this case study, lack of flexibility was very costly. The same will be true in your business.

> Beware of the trap that some companies fall into of empowering the people to initiate change for the better and then limiting what they can really do because of inflexibility of the operations systems.

Responsiveness to Provide the Timely Feedback Required for JIT

Responsiveness refers to the ability of the operations systems to process and provide information in a timely manner. JIT usually changes the meaning of "timely."

In the past, when lead times were long and inventory levels high, a period of a month or a calendar quarter was acceptable for reporting results of business decisions. With a goal of reducing cycle times and a focus for scheduling of a daily or even hourly Drumbeat, "timely" takes on a new meaning.

In a JIT environment that focuses on daily Drumbeats, information that is "batched" overnight is useless. Inventory balances, receipts of material, orders placed for raw material, and customer orders need to be updated as they become known. This requires a completely on-line system, meaning that data are processed as they are input into the system. Integrated systems, or those that "talk" to each other, are necessary to achieve this goal. For instance, if the manufacturing system does not communicate with the financial system on-line, much data retrieval will not be possible. The alternative is to have duplicate data in the different systems, which is wasteful and difficult to maintain.

Batch systems are designed with lots of built-in checks and balances that take too long to accomplish in an on-line system. These systems were designed to expect things to go wrong, with the system sorting out the errors, much the same philosophy as sorting out parts before shipping to the customer. An on-line system designed to support the JIT mindset focuses on eliminating the exceptions—solving the problems as they occur.

Determining the Basic Requirements for Systems to Support JIT

Inventory and bills of material remain pretty basic in a JIT environment. However, there are a few important requirements that your system may or may not have, so they are worth mentioning.

Making Sure the Inventory System Supports JIT

The inventory system should have multiple stores locations for parts used in many different manufacturing areas in many different products. It also needs the capability to keep inventory balances on the system for each of the locations where the common part is being used. Therefore, the number of stores locations required for the same part is relative to your products.

In an ideal JIT environment, parts are received at the dock and sent directly to the manufacturing line for use. No incoming inspection is necessary, because the parts are certified with the supplier. No extra storage is needed, because the supplier delivers the parts on the same Drumbeat that product is being built.

A few companies have lost control of their inventory by having one inventory balance for a particular part while storing batches of that same part in many locations. As people take ownership for their own inventory, they need to check that balance regularly in order to uncover early on any problem with maintaining adequate inventory. The method of having only one stores location has been termed "wall-to-wall inventory." Do not accept this idea as a viable solution for tracking inventory.

Inventory data should always be on-line to avoid limiting flexibility. The system should allow for easy flagging of those parts that are certified and no longer need inspecting. Workers should be able to receive and utilize parts in manufacturing with no additional transactions.

The issue of common parts deserves a little more discussion, since it is usual to have common parts in almost any manufacturing company. The idea of receiving parts and sending them directly to the manufacturing line (or having the line receive the parts) seems straightforward when the part being received is used only in one location and for one particular product. However, things become a bit more complex if the part is used in multiple locations for multiple products.

Separate purchase orders for each line and product produce a lot of wasted paperwork and transactions and possibly hinder your ability to get better pricing and delivery from the supplier. One company solved this problem by writing a software routine that interfaced with the business system (of course, the same thing could be accomplished manually if need be). As the common part used in different areas and products was received, the software checked the inventory balances for the part in each location against the plan for future need to allocate the right number of parts to each location. The software informed the person receiving the parts on-line how many parts to send to each location.

Of course, for this routine to work, inventory balances and the build plan must be correct; the software must be flexible enough to add such a routine and modify it as requirements change.

The Impact of JIT on the Bill of Material System

The bill of material system remains basically the same as always. However, the way different systems are structured may limit how change can occur. The bill of material system should make it easy to change from multiple levels in a bill of material for a product to a single level. You should be able to retrieve data in any format required.

The bill of material system should be able to function with a single-level bill of material. As flow is implemented, allowing product to move start to finish, a single-level bill of material more clearly reflects the manufacturing process. There is some question, then, about whether manufacturing bills of material are the same as engineering bills of material, which usually reflect how the product is designed and how each step of manufacturing is executed.

Companies have many different approaches to the documentation of product design. Many of these approaches are driven by the nature of the customers and product, so there doesn't seem to be one right answer to the question of whether to use one or two bills of material to accommodate manufacturing and engineering. One thing to remember, though, is to keep it as simple as possible and to avoid duplicating data if at all possible.

Simplifying Planning Systems With JIT

The *master schedule* is the build plan by product families that has been developed from the overall production plan. This plan takes into account marketing, orders, inventories, manufacturing capacities, and other issues that may affect the product needed and the capability to fill that need.

Traditionally, the master schedule has been organized on a monthly basis; in some companies, a quarterly schedule is emphasized. The systems have been complex, taking into account all those variables, or things that may go wrong.

In a JIT environment, the predictability of results is increased manyfold. Capacities are visible and clearly understood, suppliers deliver to the Drumbeat high-quality parts, and manufacturing builds to the daily Drumbeat.

Master scheduling is simplified to the point that it requires no more than a personal computer with spreadsheet software. Daily rates are now emphasized, rather than monthly or quarterly buckets.

The Material Requirements Planning (MRP) system typically looks at the master schedule to see what is needed to build, the bill of material to see what parts should be built or ordered from the suppliers, and the inventory balances and the purchase orders already placed to determine what is already available or on the way to decide what is required to satisfy the master schedule. All of these calculations are made in time buckets, so that parts and build plans satisfy customer demand in a timely fashion.

MRP is traditionally a complex system with many "switches" and data points available. For instance, minimum- and multiple-batch quantities can be determined; the cycle time or lead time from a supplier is considered. Traditional MRP is a batch system designed to support batch manufacturing.

As JIT is implemented, flow replaces batch manufacturing, and a regular Drumbeat replaces detailed and complex scheduling. The need for MRP is still present, but the mechanics of how the system may function will change.

Making Drumbeat and MRP Work Together

If a company is building to a JIT Drumbeat, then all regular material needs can simply be ordered from suppliers according to the Drumbeat, considering the lead time from the supplier. In other words, if ten per day is the Drumbeat, it's not difficult to order from the supplier to support ten per day. MRP calculations are not needed. However, it is difficult and may even be impractical to order every single part on a Drumbeat. Those parts that are standard, inexpensive, and not too large may be ordered in larger quantities according to MRP. In fact, most companies never eliminate the need for some type of MRP system for these types of parts.

The other use for MRP is to make calculations of needs during an exception. For instance, if an entire shipment of parts is damaged, a quick MRP may be required to calculate when you will run out of parts and how many you should order in certain time frames.

To support this requirement, MRP becomes on-line and selective. In the situation described in the preceding paragraph, you would want to run MRP only on those products using the damaged part, and you would want a quick answer in order to respond to the need.

Another use for MRP is to generate the supplier forecasts discussed in Chapter 11 describing JIT purchasing.

Frequent replanning with the flexibility to support flow manufacturing and daily or smaller buckets are musts for an MRP system to support the transition to JIT.

Taking the Waste Out of Production Control Systems

Capacity Requirements Planning is a very detailed traditional system for calculating capacity to build certain parts in a particular work center. In a traditional environment, where machines and operations are separate work centers and many different parts are built at each work center, this type of system is necessary in order to calculate what can be built within certain time buckets.

As flow is implemented and queues and batches disappear, a detailed capacity planning system is no longer needed. The capacity for a flow line is simple to calculate and easily understood. Capacity requirements planning in a JIT environment becomes more like "rough-cut capacity planning"—a more general calculation of capacities and the master schedule.

Shop-Floor Control (SFC) systems keep track of material in process by requiring transactions at specified points in the process to report the whereabouts and quantities of material in process. SFC systems usually have the capability to be quite detailed and complex. However, the detail and complexity are generally determined more by the way the data are input rather than by the system design. For instance, a particular part may have one reporting point or many, depending on how the data for reporting that part are input.

When cycle times are long, the need to know where in the process the material is may be justified. But when cycle times are reduced considerably and flow manufacturing is implemented, the only required information may be that the parts are either in manufacturing or are done. For example, if the cycle time for a part is one month, there may be some need to know where in the process the part is; if the cycle time is one day, it's in and out of the process so fast that the data are rapidly out of date.

The focus should be on developing shorter cycles rather than on implementing a complex system to track material that is in manufacturing a long time. If you don't already have SFC, don't get it! If you already have SFC, be prepared to greatly simplify or reduce the data being handled by the system. Flow and demand pull replace the need for SFC systems.

Traditional production-control systems are based on batch manufacturing through the use of work orders, or shop orders. A work order is a record in the system that is opened and closed for certain quantities of parts to be built. As parts are built against a work order, the quantity to be built decreases. Typical reasons for the use of work orders are:

- To authorize the start of work
- To communicate the steps in the manufacturing process
- To authorize/control material issues
- To track work progress
- To create a basis for cost collections and analysis
- To control material receipts
- To make engineering changes
- To determine shortages

Are work orders necessary with JIT? Flow manufacturing is done on the basis of rates, not batches. Therefore, the only valid reason to have some record in batch format may be linked to a requirement to trace product. These requirements are generally found in the defense and the medical industries; in both cases, the federal government has strict requirements on control and traceability. In some cases, these requirements can be accomplished through the use of dates rather than work orders. If work orders are considered necessary, they should be as invisible as possible.

Any system selected to support JIT should have the capability to work without work orders, focusing instead on build rates. Since the transition from a traditional environment to JIT is evolutionary, it may be necessary to have the capability to manage some product lines without work orders and others with work orders for a time. Few systems today have the capability to be "work orderless." Many systems claiming to support JIT manufacturing still have a batch record in the system, carefully camouflaged.

Simplified reporting of inventory usage and activity becomes essential as JIT is implemented. Traditionally, work orders are created and released, a parts list is made according to which stores pull all the parts necessary to satisfy the work order, and the parts are delivered to the manufacturing line after they are issued to the line through a transaction. Typically, work orders are delivered even if some parts are missing; manufacturing builds as much as possible and then waits until the parts come in.

> One of the first rules is never to deliver a kit of parts for manufacturing unless all parts are present. This will focus attention on getting the parts in time and prevent in-process inventory in manufacturing.

As JIT is implemented, parts are received and go directly to manufacturing. The flow through manufacturing is quick now, and the parts used in building the product can be deducted from the inventory balances as the product build is reported. Unnecessary transactions, or waste, are avoided.

Let's use as an example Product X. The bill of material in Figure 12-1 shows that one *A*, one *B*, and one *C* are required to build a Product X.

A typical system would produce the kit list to pull the *A*, *B*, and *C*

Figure 12-1. A bill of material for Product X.

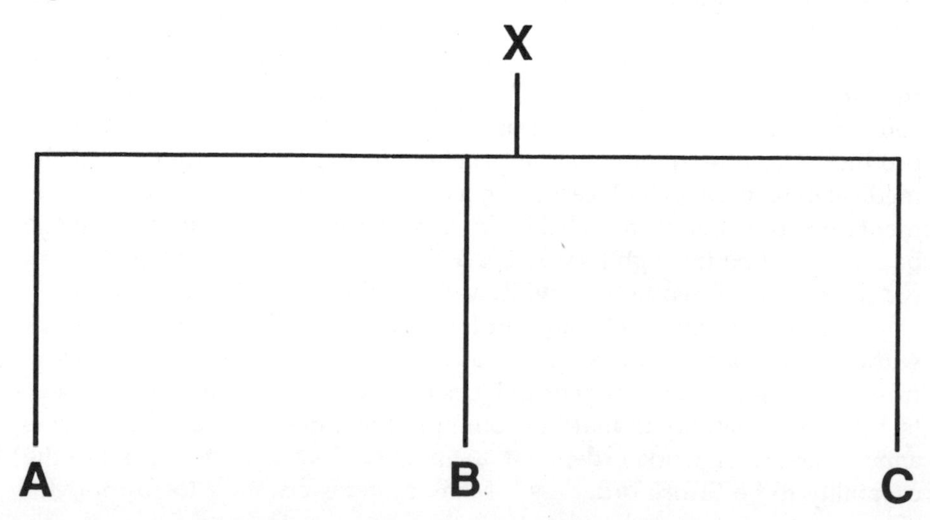

parts from stores when a work order is released to build an X. After the parts are kitted, an issue transaction would decrease the balances for each of the three parts from the stores inventory, as shown in Figure 12-2.

In a JIT environment, the parts are stored in the manufacturing area and pulled as needed by the manufacturing people (Figure 12-3). When an X is completed, a transaction called post deduct, or backflush, is done to report that an X is complete; at the same time, the inventory balances for the three parts used to build the final product are decreased accordingly. At this point, raw material and work-in-process inventory are considered one bucket, sometimes known as RIP, rather than two.

In order to complete the post-deduct transaction, an employee requires a list of all parts used to build a product in order to know which inventories to adjust. This seems simple at first glance—one can just use the bill of material. Many systems today post deduct using the bill of material against a work order, or batch record. But this design does not support JIT adequately.

If a common part is used, the bill of material will not tell the system which inventory balance to decrease. There must be a way for the system

Figure 12-2. The traditional issue of parts.

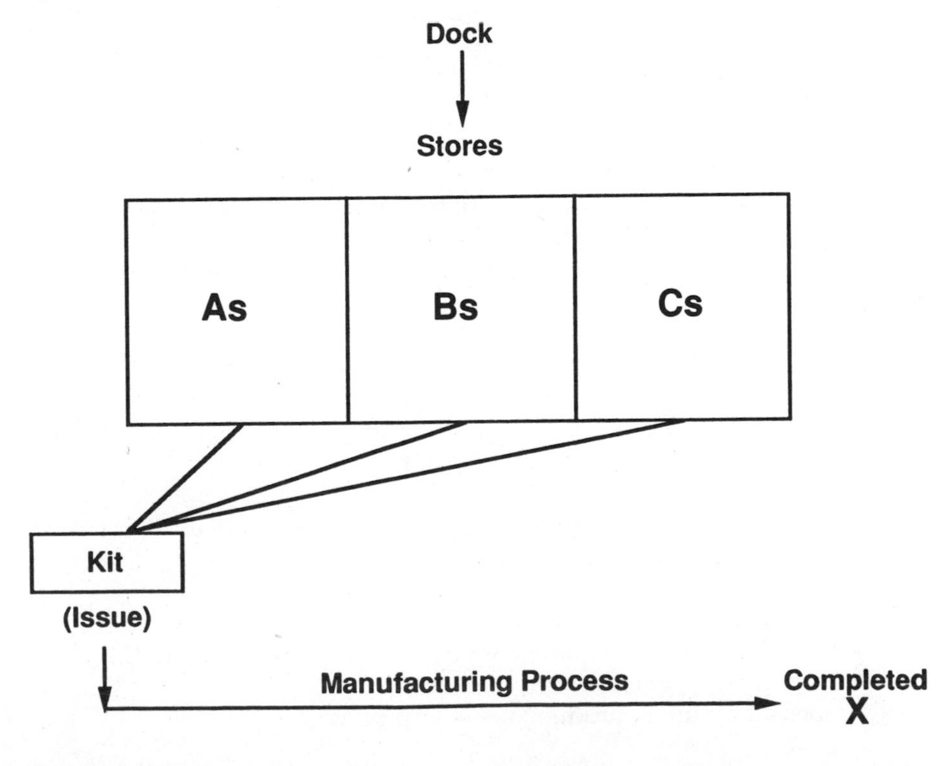

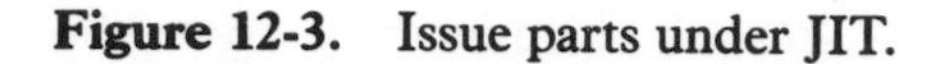
Figure 12-3. Issue parts under JIT.

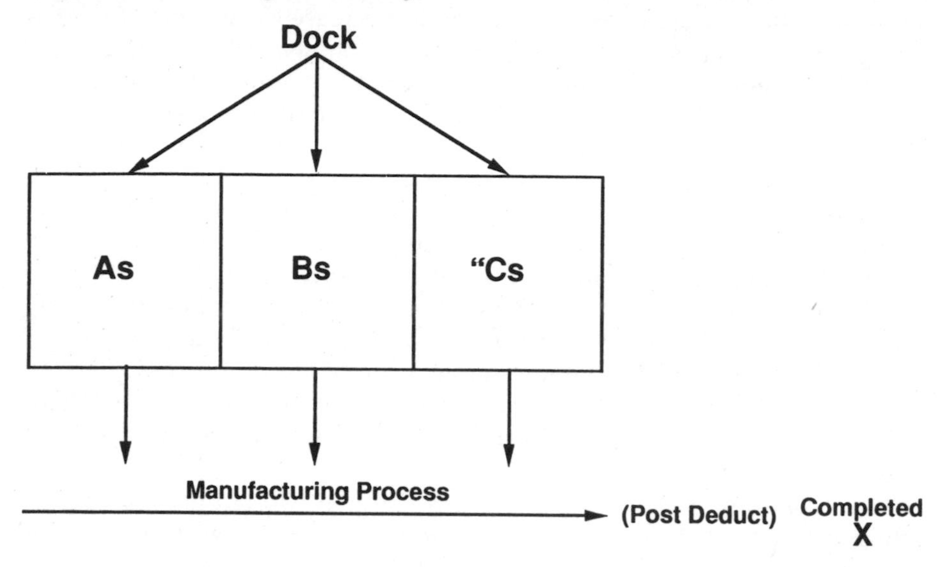

to associate common parts with certain products and manufacturing lines if it is to post deduct effectively. This may require a deduct list of some sort. The system should also be able to post deduct without a work-order record in the system, simply reporting actual production against the Drumbeat on-line.

One more consideration for post deduct: If the cycle time for a product is greater than one day, there may be a need for some intermediate deduct points to report progress and decrease inventory balances for parts already used. Unless you have no longer-cycle products, this is a feature you should require when and if you decide that a true JIT-compatible system is required.

Figure 12-4 summarizes the changes in production-control systems by comparing in a macro sense the traditional batch with JIT.

There is little impact from JIT on financial systems, such as general ledger and accounts payable. However, costing systems can change significantly. This issue is discussed in more detail in Chapter 14.

In summary, a business system designed to support a JIT environment is based on the following assumptions:

- Flow manufacturing is implemented.
- Ownership and discipline are built into the environment.
- In-process inventory is reduced dramatically.
- Focus is on linear production—a Drumbeat.

Figure 12-4. As JIT is implemented, the systems are simplified.

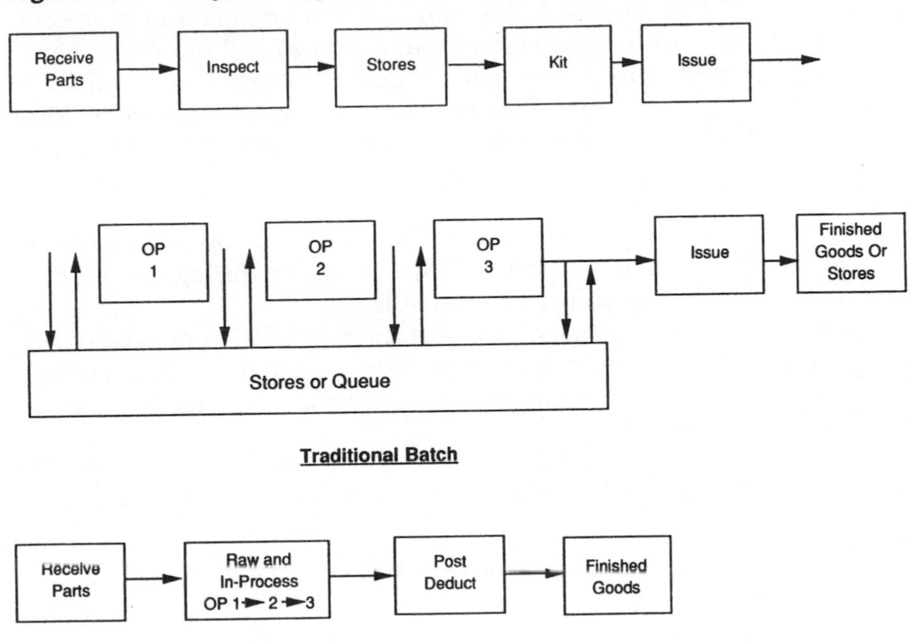

Tools to Consider for JIT

Personal computers with the ability to retrieve and send data to the main computer can increase productivity. A few simple programs, including spreadsheet technology, are required.

Another tool that is often discussed is bar coding. Executives at many companies have mentioned bar coding to me as a way to automate shop floor control, or the tracking of detailed labor data. Using bar coding in this manner is automating wasteful processes! Eliminate the need for all this data, and you won't need all those terminals and bar-coding devices.

Bar coding does have a great deal of value when utilized in a different way, however. Bar coding eliminates errors and speeds up the process of collecting data. Bar coding is ideal for such transactions as receipt of material or post deducting. Automating for the right reasons makes sense.

How Traditional Systems Can Work as JIT Is Implemented

Many companies have been successful in implementing JIT with traditional batch systems. Much can be accomplished just by changing the data within the system without changing the functionality of the system itself; for instance, instead of tracking in and out of every operation in a process, structure the data so that the system tracks only in and out of the entire process, or manufacturing flow line. Shop-floor control may still be in place, but such a change will eliminate many wasteful transactions. Most companies that have used traditional batch systems find that as they implement JIT they use only a very small percentage of the system's total capabilities—as little as 10 percent in some cases.

The transition from traditional to JIT must also be considered. A system should be able to support both environments during implementation, which may take several years to complete.

The timing and requirements for operations systems should be understood early in a JIT implementation in order to avoid needless systems purchases or changes. Action to change the operations systems may occur later in the implementation process.

The most important considerations for operations systems are that the system must always support the way business is done and must never inhibit progress with JIT.

Chapter 13
Improving Product Design

The process of product design, as well as the design itself, can have a significant impact on your ability to meet customer requirements. JIT principles have typically first been applied to manufacturing processes, gaining solid results in a relatively short period of time; however, the product-design group should also be involved in any JIT implementation from the outset to avoid missing the longer-term high impact of JIT. Without the involvement of product-design engineers, only short-term results will be achieved with a JIT effort. Permanent change in the way a company manufactures products will not occur without the involvement and commitment of the product-design group.

JIT can offer benefits to product-design engineers by providing better teamwork, more timely feedback, and measurements that make sense for success. The establishment of a common company vision focused on meeting customer requirements will drive common goals for manufacturing, marketing, and engineering, eliminating some of the barriers that often divide these important functions of the organization and encouraging teamwork instead. I believe that these barriers developed not because any one function desired them but because conflicting departmental goals and measurements forced their creation in order for the individual departments to be successful.

What design engineers like best is designing products; in fact, that is their all-important contribution to the company. Successfully implemented JIT principles can free up design engineers to spend more time and energy designing better products that are more easily manufactured to meet customer requirements.

The way a product is designed has a significant impact on the way the product can be built and whether it supports new goals such as flow, one-at-a-time, balance, and flexibility. Product design also has an impact on the quality levels achieved in manufacturing, and the sourcing of material during the design phase can help or hinder JIT purchasing goals.

This chapter is not about how to design products—that varies greatly among industries. Instead, the focus is on a new mindset for, and approach to, product design. First, I discuss the changes in requirements for success in product design within a JIT environment and how to achieve success under these new requirements. I also discuss how to improve the "time-to-market" window—the time from design to release—for new designs and ways multifunctional teams and parallel efforts can make it easier to work within those windows. Finally, the benefits of this new way of operating for product-design professionals are outlined.

The Impact of Traditional Goals and Measurements for Product Design on Overall Success

Traditionally, product-design engineers are given three basic goals for a new product: (1) functionality; (2) a budget for direct costs of material, labor, and overhead; and (3) a schedule. Let's discuss how each of these goals affects the resulting design.

The Impact of a Focus on Functionality on the Resulting Product

What is often missing from the traditional design process is a carefully throught-out functional definition of what the product should be. Without a clear product definition, usefulness has been defined more in the mind of the creator (design engineer) than in the mind of the customer; consequently, overdesign has been common.

Who Is the Customer?

I was invited to a high-level marketing meeting about a new product that was before its time technically and that had taken years to develop. I was there because I had found an application for the product in the plant. The topics of discussion amazed me: What market might the product fit into? Who might buy it? What are its attributes and selling points? It was clear that the engineers had been working in a vacuum, unaware of customers' needs or desires. Engineers had designed a product they assumed the market would love. Now that the product was done, marketing was trying to think of who would really love it!

The ending to this story was that sales volume was not ever very high on this particular product. However, the company was lucky in that as a result of the technology breakthroughs made in the development process, other products were developed that have dominated the market ever since.

Functionality has typically also been judged by engineering. "It works" means that engineers built one prototype in the lab, sourcing all parts with no problems. Because of the way that functions are organized and measured, manufacturing and marketing do not usually judge whether "it works" until engineers have handed the new product over to them.

The Role of Budgets in Driving Product Development

Direct costs of material, labor, and overhead usually make up the complete cost goal for engineering purposes; the ability of engineers to stay within the cost goals is a primary measure of their success. For instance, there are many instances of engineers designing a manual operation that introduces a great deal of variability into the manufacturing process rather than using a more expensive part from a supplier. This meets the engineer's objective of keeping direct material costs down, but the cost of rework and scrap in manufacturing because of the variability of the process far outruns the cost of the supplier part. However, the engineer, having met the goals set forth, is not penalized for, nor is he or she probably even aware of, the additional quality cost to manufacturing.

Two Sets of Rules

Burn in is a term, used in any type of electronic product, that means letting the product run for a while to see if any of the electronic components fail early in their life.

One company manufacturing printers had implemented flow and balance into its manufacturing processes. Its queues were approaching a quantity of one in most locations. But a new product was introduced with a two-day burn-in cycle; so in the midst of a very nice flow sat two days' worth of inventory.

The engineers were judged successful, since this queue did not affect the direct labor, material, or overhead (based on labor, of course) for the product. It was as if engineering was working to one set of rules and manufacturing to another.

As should be clear by now, direct material, labor, and overhead represent only a piece of the total cost of a product. When engineers are required to focus only on these portions of the cost, they are destined to make decisions that are not correct for the good of the entire company.

The Impact of Old Assumptions About the Design Process Schedule on Success

Traditionally, a schedule for the design-to-release of the product is agreed to but almost never taken seriously or adhered to by anyone in the company. The assumption is that it is impossible to schedule a creative endeavor; new products are ready when they get ready. In fact, the assumption by all concerned is that release of new products will always be late; if a product release is on time, no one in manufacturing and marketing will be ready anyway.

Focusing Product Design on the Customer

JIT drives a focus on meeting customer requirements (both internal and external) with the least amount of cost. As this common vision is established for the company and extended to product design, the goals and measurements move away from the traditional ones to the following requirements (Figure 13-1):

- 100 percent quality design in order to support 100 percent quality the first time in manufacturing
- Cost objectives that include total cost of the product through manufacturing and to the customer
- Timeliness—schedule adherence to meet the market window
- Performance and design requirements that meet (don't beat) customer requirements

After developing a clear vision of how to meet each requirement, I will move on to the elements and basic assumptions necessary to help product design achieve success in supporting the company vision.

Designing Quality In

Product quality within the manufacturing process can be affected significantly by the product design.

Figure 13-1. Meeting customer requirements with least cost.

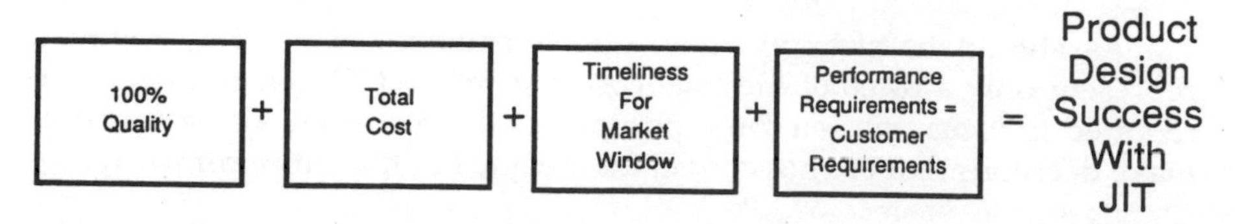

Designing Quality In

In one case, a certain product was designed with a two-piece plastic cover requiring four bolts for assembly. The pieces could actually fit together either of two ways; however, only one way was correct. Mistakes were easily made in manufacturing.

The people on that line suggested that a snap-together cover be developed—one that snapped only if the two pieces were assembled correctly. Not only was assembly time saved, but it was impossible to get it wrong. Sometimes, this is referred to as a fail safing—making sure the operation can only be done correctly through the design of the product.

Product engineers can gain information on the impact of design on quality by "walking in manufacturing's shoes." The more data they have about what it's like to manufacture the products they are designing, the more successful they will be at designing products that can be produced with 100 percent quality the first time.

Establishing better communication links between manufacturing and design is imperative. Productive feedback loops develop as people have a chance to know more about the represented processes. Forming product line teams with all functions represented is a good place to begin.

Do you remember that "I am" statement from Chapter 4 on quality? "I am" includes product design. Quality can be designed in.

Using Total Cost as a Product Design Goal

Cost goals in JIT reflect total cost, not just direct cost. Total cost in the previous case study included the cost of reworking the covers that were assembled incorrectly, as well as the warranty cost for any that were not corrected and were shipped to customers.

The cost of quality includes rework, scrap, replacement parts, wasteful movement, and waiting—any cost incurred because the product was not made with 100 percent quality the first time.

Other costs that can be eliminated by careful design are those associated with flow and balance disruptions. Think back to the example in this chapter about printers. Two days of extra queue were required to satisfy the design requirement to burn in each unit for two days. Not only was space tied up, but flexibility was decreased considerably. Costly labor was also required to manage such a massive number of printers—not to mention the dollars tied up in inventory.

Frustration of the people in manufacturing is a commonly overlooked cost in manufacturing. If a design is difficult to build despite some obvious ways that the product could have been designed that would have made manufacturing easier, people become frustrated. Frustration manifests itself in poor quality, lack of effectiveness, and a general "don't care" attitude.

In order to set cost goals, the target competitive price should be determined by marketing. Once the margin, or profit, has been applied, the target cost can be determined. If product engineers have a cost goal that reflects total cost, they have an opportunity to broaden their focus on cost, focusing on the customer.

Meeting the Market Window Through Better Scheduling

Timeliness, or schedule adherence, of product design is crucial in today's competitive marketplace. The flow of creativity can no longer determine schedules for design; schedules should allow the right amount of time for design and manufacture of the product to meet the marketing window, the time in which the product will be successful in the marketplace.

Often, unforeseen problems make the design cycle take longer than expected, but the market window stays the same. Manufacturing is then pressured to manufacture a new product in record-breaking time, causing more problems to surface. Later in this chapter, I discuss how total time can be compressed without compromising design.

Meeting Customer Requirements for Design and Functionality

Performance requirements include designing the right product as well as emphasizing the requirements of the customer, such as reasonable price, reliability, and on-time delivery. Design can affect all of these issues; if marketing passes along these requirements to product-design engineers early in the process, the opportunity to meet the customer requirements is increased manyfold.

It is worthwhile to do a customer-needs survey if there is any chance that those needs are not well understood. Perhaps this could be a team effort involving not only marketing but also product engineering and manufacturing. For instance, a company may think that price is a major requirement for most customers. However, when the customers are surveyed, they cite faster and more responsive deliveries as being much more important as long as the price is reasonable. We all tend to make assumptions about what the customer wants—instead of asking the customer!

"Fitness for use" is a term frequently used to denote a product that meets the customer's product requirements. Exceeding those needs with technical elegance or more bells and whistles does not guarantee that a company can sell more of a product or sell it at a higher price. In fact, I have seen many instances when beating customer requirements resulted in a higher cost but not a higher sales price. The margin suffered instead.

Another requirement that is often forgotten, although it is very important in many instances, is that of "maintainability." Once the product is purchased by the customer, how easy is it to maintain? A client recently received a piece of equipment that required rework at the customer site in order for regular maintenance to be performed! It was physically impossible otherwise to get to some of the parts inside the equipment that required regular maintenance. And what about those electronic products that require a service rep to replace a printed circuit board? These products could be designed to allow easy removal and replacement of the boards by the customer, saving the supplying company significant dollars.

A good working relationship with the customer will assist you in determining how to meet customer requirements. A "both-win" partnership with customers can be just as beneficial to you as a "both-win" partnership with suppliers.

Meeting Customer Requirements Through Designing for Manufacturability and Material Sourcing

To meet the customer requirements for success in product design, a focus on two elements is necessary (Figure 13-2): designing for manufacturability and material sourcing.

Designing Products to Support JIT Manufacturing Processes

Manufacturing processes are affected by the product design. Product engineers must take manufacturing needs into account as the focus moves to flow and building product as the customer needs it without piles of inventory. As the impact of JIT changes is clarified for your particular processes, product design can help make them more successful.

Figure 13-2. Necessary elements to meet customer requirements.

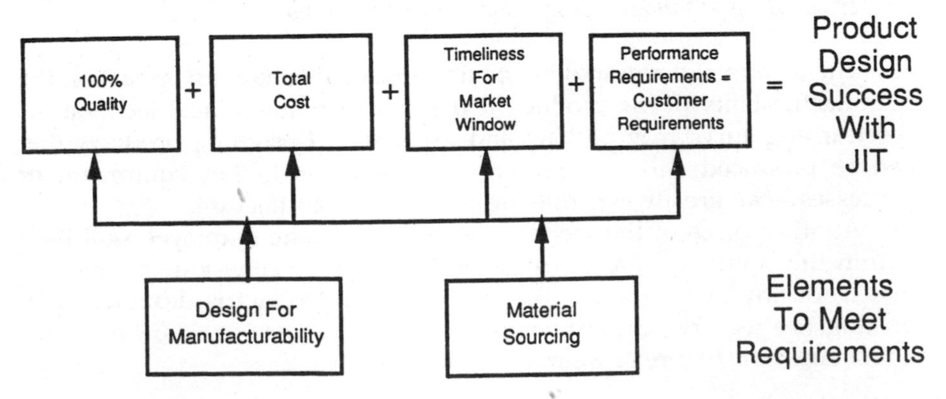

Product Design Can Prevent Flow

Several years ago, a client asked for help with a production line that built small missile casings. The client complained that the cycle time was too long, there was too much inventory on the floor, and the flow was unmanageable. Top management suggested that a team be formed in manufacturing to improve the flow and balance of the line and to create the ability to achieve a daily Drumbeat.

After a close look at the process, I discovered that after the fiberglass tubes were made, all fittings were glued onto this tube with a glue requiring hours to cure. In fact, only about 5 percent of the total cycle time was spent actually placing the fittings on the tube—95 percent of the time was wasted waiting for the glue to cure in various different stages and waiting in queue. Manufacturing had very little control over the flow and balance problems. A design change would be required to find a glueing process that did not require so much cure time.

After this was brought to management's attention, engineering was asked to work on the problem. A better glueing process was found, and then manufacturing began work on the flow and balance of the line.

Controlling Product Variation to Simplify Manufacturing

Product variation can also have an impact on manufacturability. It's always easier to make the same thing over and over than to make many different things, one after the other. The way the variation is designed into the product can affect success in manufacturing. Having product variations at every step in the entire manufacturing process is very different from building the same product up to a single configuration step or split-off point that determines which variation is built.

Setup requirements can also be reduced or eliminated by the design of the variation, allowing better flexibility.

Evaluating Technology Requirements in the Design Phase

The technology required by the design can also have an impact on the manufacturability of the product. If the design requires new technology, the learning curve may be long and expensive. Designing products that can be produced with the present technology—whether equipment or processes—can greatly expedite the successful manufacture.

Another piece of the technology required is the employee skill level within the company. Product designers should be conscious of the skill match for any new product. For instance, if a product is designed with brazing as a requirement but no one in the company knows how to braze, it will take time to hire or train the right people.

Making Inspection Easy Through Product Design

Another consideration is the need to verify quality. How easy is it to inspect the product—whether it's the next operator or an inspector doing the inspection?

Designing for Inspectability

Recently, I attended a top management meeting at a client company. The subject at hand was the fact that a product had been designed so that only a certain machine could verify the quality of the part. (Since this company makes medical products, it is essential that the product meet precise specifications.) There are only a few of these special machines in the plant, which means that workers must walk to get the part inspected and wait for the inspection. We wondered why future products couldn't be designed so that the operator could do a preliminary inspection at the work station.

When I asked the quality manager what seemed an obvious question—"Aren't you involved in design review meetings?"—the answer was that he had never been involved and usually didn't know what it was he was to inspect until product release. It was always a race to get gauges and programs to inspect the part in time.

This management team was able to get quality more involved in the product design so that inspectability could be verified as the design was developed.

Sourcing Material During Design to Support JIT Manufacturing

Material sourcing for new products has typically been done by the engineering department. Purchasing does not get involved until the product is released to manufacturing. But at least two problems crop up when selection is done by engineering alone:

1. Engineers often select a new supplier when a well-established supplier could have supplied the part, or they select a supplier who is being eliminated through a supplier reduction process. These are the primary reasons that most companies have so many suppliers.

2. The supplier selected by engineering may have good capability to support development volumes but may not be capable of supplying production quantities. In other words, business evaluations are forgotten in the midst of worrying about technical issues.

A better approach to the material sourcing effort is to involve a product-design team that includes not only the product-design engineers but also purchasing, quality, and finance. The team approach ensures that the supplier selection is sound from all standpoints. Each supplier selected for a new-product design should either be one of the "right few" as described in Chapter 11 on JIT purchasing or have the potential to become one. You can use the same selection criteria as those outlined for the selection of the right few in Chapter 11.

Three Principles for Success in Meeting Customer Design Requirements

Three principles should govern all design efforts in order to design successfully for manufacturability and material sourcing, the key elements in meeting customer requirements.

1. Simplicity of design
2. Balance between standardization and creativity
3. Stability of design

Figure 13-3 shows how these basic principles support the process of meeting customer requirements.

Making Product Design Successful and Challenging Through Simplicity

Seeking simplicity of design does not mean that the design should lack interest or challenge. In fact, a simple design can be more challenging to create than a complex one. The fewer the parts and the easier they are to assemble, the more successful a design will be. In addition, diagnosis of problems and regular maintenance become easier as the design is simplified.

Using Standardization to Enhance Creativity

Standardization does not necessarily limit engineers' creativity. Standardization of common items, such as screws, nuts, and bolts, is desirable so that engineering time can be devoted to the creation of new, leading-edge products.

Some companies have produced a handbook of standard parts with the understanding that if a part in the handbook can fill a functional need, that part is to be used instead of a new or different one. This handbook makes it easy for an engineer to check for standard parts and go on to more interesting tasks. In this way, creativity is not stifled but instead is guided and managed in an effective way.

Figure 13-3. Basic principles necessary to support the elements for meeting customer requirements.

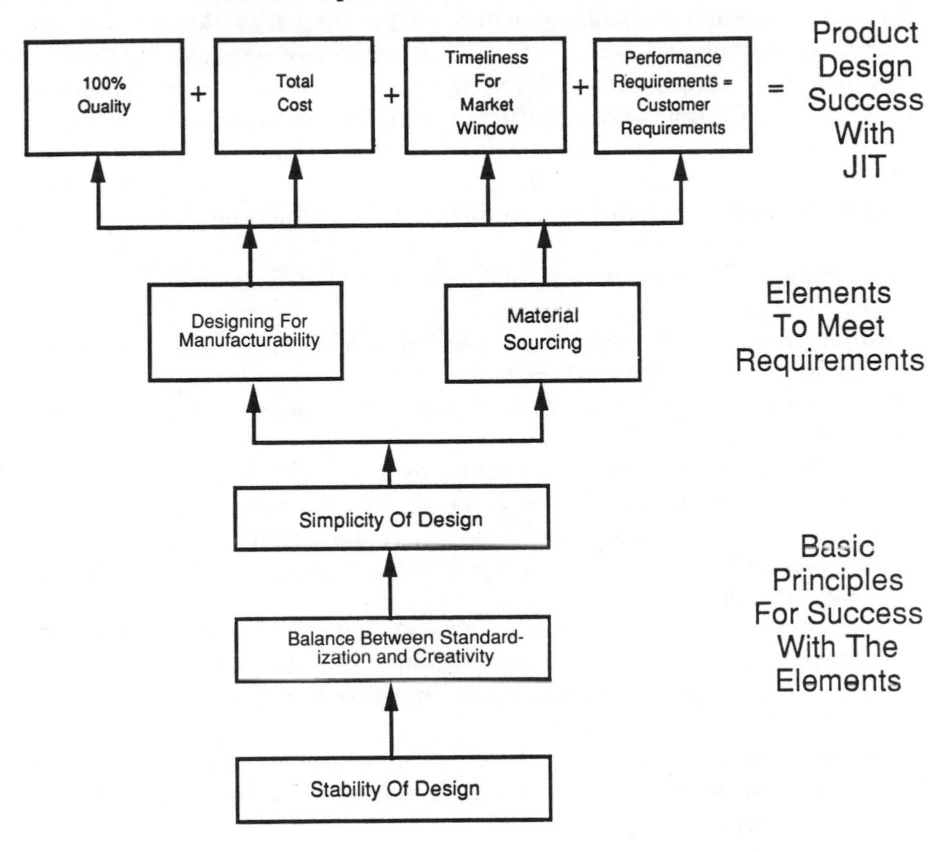

Designing in a Stable Product Design to Support Manufacturing and Materials

Stability of design is often ignored in the development phase; the focus is on getting the design workable, not making it last for the life of the product. As schedules are managed and adhered to in a JIT environment, there is more time to focus on designing in a stable product design for the life of the product. A measurement that can be imposed on product design is to count the engineering changes on each product developed. This creates a visible report that emphasizes the need for stable designs. Engineering changes are costly and disruptive and can have a detrimental effect on manufacturability and material sourcing for the product. Again, a team approach can ensure the stability of the design.

How to Meet the Time-to-Market Challenge

The "time-to-market" challenge refers to the time that elapses between design and release of a new product. The window of opportunity for new products is getting shorter all the time. Success for many companies depends on designing and introducing complex products within a certain time window; if the market window is missed, the business may be completely lost. This is especially true in the high-tech companies where product life cycles are short and getting shorter all the time.

Why Using a Linear Approach for Product Design Fails

Traditionally, products have been designed in a linear fashion—one function and one step at a time. There is the classic story of the engineer in the tower who pushes designs off the edge when done, just in time for manufacturing to "catch" them.

In Figure 13-4, the design is pitched over the fence between each step and function. Communication is blocked between functions, making it very difficult to backtrack if problems occur. Generally, there is a lot of finger pointing and confusion when problems do occur, along with significant wasted time.

Again, the linear approach with "fences," or communication barriers, developed because of a lack of a common vision and common goals. I believe that most people would agree that this is not a common-sense approach to the design process.

Exhibit 13-1, at the end of the chapter, is a study undertaken by a high-tech company to determine why it missed a market window. The

Figure 13-4. The linear approach to product design.

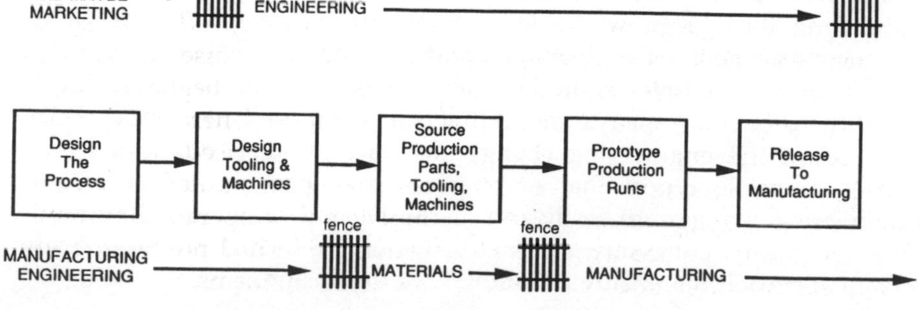

different linear steps were evaluated through the use of flow analysis and value-added analysis; conclusions about why the company slipped up followed.

You may recognize some of the conclusions:

Successes

- Product met all design requirements. There were a variety of design requirements for the product. Among them were inclusion of Unix, retention of level of performance from previous versions, and a full—not partial—implementation of the combination of operating systems. All of the design requirements were met.

Problems

- Lack of a business plan caused confusion among the different organizations. The lack of a business plan for the product caused confusion about the roles and responsibilities of the different organizations involved. Some people felt that a business plan would have shown the need for a better understanding of everyone's roles.
- The dependencies of the deliverables were not fully understood. The base level on the product was not frozen for the rest of the deliverables to build on. People thought that there were too many changes to permit accurate resolution of problems.
- The schedule was not realistic and was committed to too early in the process. The schedule was felt to be unrealistic. The team felt it did not have input into the formulation of the schedule. People on the team expressed the feeling that they had to commit to the schedule too soon, before they fully understood the nuances of the project.
- Not everyone on the project knew Unix well. Some people on the team had to be trained to use and understand the Unix operating system. People felt that if everyone knew Unix, there would have been fewer problems merging the two operating systems.
- As earlier dates slipped, beta end date was not revised.
- There were too many beta sites (approximately 180). While the beta test was well-planned, there were problems because of slipped dates. As dates for earlier stages were missed, the beginning date for beta was slipped, but the end date was not. There were more beta sites than normal, and some sites were not serviced adequately.
- Customer release was delayed as a result of a new, untested release method. Along with releasing a completely new operating system, the company used a new release method that had not been adequately tested and that initially caused some problems.

Goal Conflict

Recently, I was working with a manufacturing team that was experiencing problems with a new-product design. The group had some ideas for engineering about how to improve success in manufacturing by changing the design.

The product engineers were aghast that the manufacturing people would question their design—it met all traditional requirements, and the prototype built in the lab worked just fine. The supplier selected by engineering had no problem supporting the development quantities of parts.

It was pretty evident that the goals of the two groups were not the same or even understood by one another.

Using Teamwork and Parallel Steps to Achieve a Competitive Advantage

A new and very successful approach to product design incorporates two principles:

1. A multifunctional team gets more done better and faster.
2. Some steps in the usual product-design cycle can be made parallel to shorten the cycle and provide quicker feedback.

Notice that in Figure 13-5, marketing, engineering, manufacturing engineering, materials, suppliers, manufacturing, quality, and finance people are included on the product-design team. You may include other functions, as well. The key is to include everyone who has a stake in the new design. Such a diverse team generates more and better ideas, identifies and solves problems more quickly, and improves communication between steps. The team makes better design decisions that can have an impact on the overall success of the new product.

Parallel efforts not only shorten the cycle time but also have other advantages, including quicker focus, better feedback, and speedier problem resolution.

The most outstanding advantage of the team and parallel approach is that of ownership. Involving many functions early in the product design helps develop ownership, which leads to commitment to the success of the new product.

A Success Story

When we began the JIT implementation at Valleylab, engineers didn't seem to understand how any of the changes might involve them. It was difficult to get their interest and support. While they were tentative, they were also receptive. They did attend some training, and we kept talking.

Figure 13-5. Using teamwork and parallel steps to meet the time-to-market window.

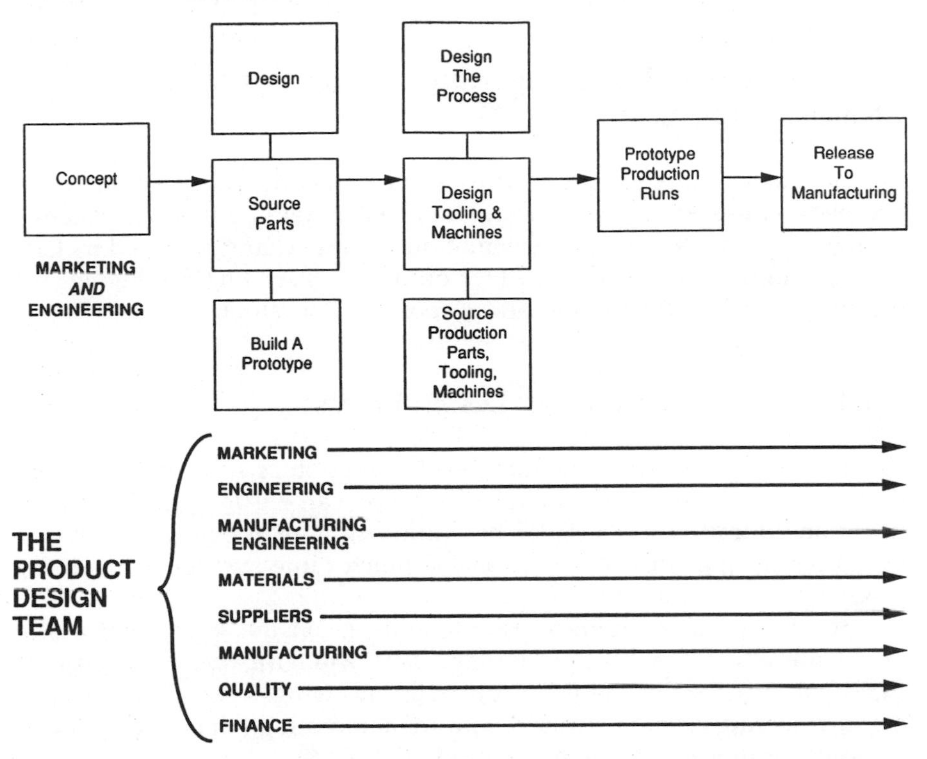

During one visit, I noticed that manufacturing was beginning to become a resource for engineering. In fact, the technicians in manufacturing had identified a problem and had worked with engineering to help diagnose the cause of the problem and to find a solution. Engineering began to see the JIT effort differently—instead of being just one more thing to deal with, it became a new resource. If the manufacturing people could help engineering identify and solve problems, engineering would have more time to work on new products, incorporating learning as it went.

Today, the ideal is taking place in the design of a next generation surgical-pencil generator. The multifunctional project team is working on the new design. Engineering is making a special effort to understand the vision for the future in manufacturing and to design the product to support that vision. There's no question—that new product will be a winner!

Applying JIT Principles to Process Design

Most of this chapter has focused on new-product development, but the principles discussed apply to process design as well. Process design in a

JIT environment is done with a team made up of people who will be manufacturing the product, as well as product engineers.

How JIT Product Design Can Apply to Engineering Change Management

Engineering changes should decrease significantly the JIT implementation proceeds. However, in the real world there will always be some engineering changes. The design and implementation of these changes should include the same ideas as those discussed for product design. Engineering-change management should be simple and responsive to customer needs.

Applying JIT to the Product Design Process

Creating awareness and education are the first steps in getting product-design engineers involved in a JIT effort. You can ensure that this awareness and education focus on JIT principles from an engineering perspective. I have found that this emphasis is much more successful in getting product design involved.

Second, involve product-design engineers on the teams that are to guide the effort from the beginning. This will give them a chance to experience the results and get involved in the changes early. Manufacturing engineering often is involved with manufacturing pilots as a technical resource for those teams; why not product design?

Third, product-design engineers can begin to apply the JIT mindset to their own process of design as a way to begin to make the transition from linear design steps to parallel steps. A value-added analysis can be performed in product design as a way of documenting their processes and selecting improvement efforts. Documentation of the linear steps that take place in a product-design cycle through the use of flow analysis can be a first step in determining how to begin a team with parallel efforts. An example of one client company's documentation is shown in Exhibit 13-2 at the end of the chapter. This list helped identify opportunities for improvement.

Four Benefits of JIT for Product Design

Product design can benefit from a JIT implementation in several ways:

1. More and better ideas result from the team approach.
2. More time is freed up to focus on new designs and less time is spent fighting fires.

3. Problems are identified and solved more quickly and more easily at lower cost.
4. Product-design engineers are seen as valuable team members; communication with other functions is facilitated.

For long-lasting results with JIT, product design is critical. Don't delay in getting people from this function involved; it is a long-term process, so the sooner you begin, the sooner the results will be evident.

Exhibit 13-1. Missing the market window.

WORK FLOW HIGH-TECH PRODUCT

STEPS	Degree of Team's Experience With Task (High / Medium / Low)	Time Required Actual vs. Expected (Less / Same / More)	Cause of Variance
0. Decision			
1. Decide to "do product"			
I Product Definition	L	S	
2. Team formed			
3. Planning			
4. Plan review			
5. Target product calendar			
II Hi-Level Design and Planning	L	M	Schedule not realistic, not proactive. Little involvement
6. High-level functional specification			
7. Integration schedule			
8. Preliminary plans			
9. Test planning			
10. Product calendar target update			
III Detailed Design and Committed Plans	H		
11. Design notes			
12. Functional specification			
13. Committed final plans			
14. Committed to product calendar			
IV Implementation and Early Test	L	M	Merging two systems; radical change; size of release; lack of knowledge of Unix, Complex; interdependencies not understood; inaccurate forecast
15. Marketing plans			
16. Early announcement			
17. Software development			
18. Documentation			
19. Beta test preparation			
20. Customer service forecast & internal plan update			
21. Alpha test	M	M	Lack of planning
V Field Test and Quality Verification	L	M	Beta gets squeezed, Schedule drives end date
22. Beta test			
23. Internal product qualification			
24. Customer release preparation	L	M	New release method never tested
VI Customer Release	H	M	Never had a business plan; communications with marketing not clear; relationship with customer service not helpful
25. Software			
26. Documentation	L		
27. Post evaluation			
VII Continuing Product Support			
28. N/A			
VIII Retirement			

Exhibit 13-2. Linear steps in product design.

A CLIENT COMPANY'S PRODUCT DESIGN STEPS

STEP	WHO	CUMULATIVE MONTHS
1. Charter Established	DESIGN, MKT, SLS	1
2. Sketches Drawn	DESIGN (DSN)	2
3. Select Sketches	DSN, MKT, SLS	2
4. Build Prototypes	DSN, PROTOSHOP (PS)	5
5. Cost Product	R&D, IE	5
6a. Critique	DSN, MKT, SLS	5
6b. Preview at Mkt	DSN, MKT, SLS	6
7. *Focus Group (Style, Line)	DSN, MKT	6
8. Design and Select	DSN, MKT, SLS	6
9. Cost Product	R&D, IE	6
10. Develop Materials	DSN, MKT	9
11. Cost Product		10
12. *Focus Group (colors, materials)	DSN, MKT	11

* Atypical Events

A CLIENT COMPANY'S PRODUCT DESIGN STEPS
(continued)

	STEP	WHO	CUMULATIVE MONTHS
13.	Build Prototypes	DSN, PS	12
14.	Cost Product	R&D, IE	12
15.	Show	DSN, ALL MGMT	12
16a.	Design Added Features	DSN, MKT, SLS	13
16b.	Change Material Specs	DSN, MKT, SLS	13
17.	Alter Vendor Process	DSN	13
18.	Cost Product	IE	13
19.	Teach Shop to Build	DSN, PS, MFG	14
20.	Distribute Work	MFG, DSN, PS	14
21.	Pilot Run (Test)	MFG, DSN, PS	14
22.	Consider Design/ Manufacturing Issues	DSN, PS, R&D, IE MFG	15
23.	Cost Product	R&D, IE	15
24.	Design Simplified	DSN, MKT, SLS	15
25.	Sold Into the Trade	SLS	15

Chapter 14

Measuring Performance to Drive the Change to JIT

Corporate America is a success-oriented culture. Often, the grades on the report card are emphasized more than the content of the subjects graded. JIT forces management and employees to look at the content of the work subjects they are graded on.

There are two types of measurements—performance measurements to drive change and measurements to report what is happening today. This chapter focuses on the right performance measurements to drive the change process associated with the implementation of JIT.

Following a discussion of each of twelve financial indicators and how they change for JIT, some innovative costing models are outlined. Next, I focus on people measurements that make sense for JIT.

In this chapter, I discuss ideal performance measurements. Of course, you can't change them unilaterally, but the more feel you have for what good performance measurements look like and how the present performance measurements hinder JIT, the more persuasive you can be when you argue for a change.

Measure the Right Things to Drive Improvement

The performance measurements selected should be those that most effectively highlight problems, not those that make people look the best.

Looking Bad Can Be Good

A client recently shared a story about a meeting that he was asked to attend. The other attendees were one of his plant managers, his controller, and a top industrial engineer. The subject was a new corporate edict that there would be no more loss allowances.

The controller said, "I can perform some accounting tricks so we look good."

The industrial engineer said, "We can change the standards so we look good."

The plant manager almost buckled but stuck to it and said, "I don't want to look good. I want to see how bad we look so we know what we have to work on to get better." That was a victory! This plant manager has the right idea about the only valid reason for measuring.

When implementing JIT, having the right performance measurements—both for financial data and for people—at the right time is as important as having the right parts at the right time. Business decisions are made primarily on financial measurements; if the wrong things are measured, the wrong decisions will be made. The measurements for people define success in an organizational sense; if the wrong things are measured, people will work on the wrong things in the wrong way.

Therefore, performance measurements—for both financial and people—should drive your company toward the vision for the future. As in all facets of JIT, *simple* and *visible* are watchwords. Detailed measurements focusing on micro issues are replaced with macro measurements that reflect the company's true health and growth. The challenge is to pick the right few measurements that, when focused on, provide more detailed results; these right few measurements serve as vital signs for your company.

Driving the Change With Twelve Financial Indicators

There are twelve financial measurement categories that have prevailed over the years in manufacturing companies. Each can be changed to drive the change to JIT. Some of these changes can be implemented as pilots in the early stages of a JIT implementation, perhaps requiring duplicate measurement systems until all areas of the company are applying JIT principles.

1. Focusing on the Customer for Schedule Attainment

EFFICIENCIES/UTILIZATIONS → DRUMBEAT/LINEARITY

Drumbeat refers to a regular, linear schedule. Daily or hourly rates are the focus, not monthly or quarterly goals. Only 100 percent quality product is considered in measuring Drumbeat attainment.

A focus on linearity instead of just "making the numbers" will cause many good things to happen, including increased quality levels, less work-in-process inventory, and better cycle times.

Methods for measuring Drumbeat have been discussed; the keys are to use the Drumbeat charts to expose problems and to make sure that the results of this measurement are clearly visible in the plant.

Drumbeat should replace the traditional measurements of efficiencies and utilizations. Efficiencies have typically measured the quantity of parts produced per labor hour; more has always been considered better. Efficiencies are often utilized to evaluate an employee's individual performance.

Utilizations measure equipment utilization rather than the utilization of people's effort; running the equipment as fast as possible as many hours as possible yields good utilization marks.

The principle that these measurements reflect is "More is better." Keeping people and machines busy is better than having them idle, even if they are producing the wrong goods or goods that are not needed at that time—goods that are going right to inventory.

Efficiencies and utilizations inevitably suffer as JIT is implemented. After all, people and machines cannot build like crazy anymore; they can build only when something is needed, even if it means stopping temporarily when no product is needed. If the measurement is not changed, you can imagine the mixed message workers get, as if someone were saying, "Your efficiencies must be higher to get that raise, but you are no longer allowed to build unless more product is needed."

An emphasis on Drumbeat, or linearity, focuses the team on building the right thing at the right time with 100 percent quality to satisfy the customer need.

2. Providing the Internal Customer 100% Quality to Ensure 100% Quality to the External Customer

EXTERNAL CUSTOMER → INTERNAL CUSTOMER

Quality measurements in a JIT environment focus on "never building a bad one" rather than "never shipping a bad one." Watch defect rates, scrap rates, and rework data on a *daily* or *hourly* basis—not just at month's end. This requires that data be on-line and visible. Statistical process control can provide warning data that let you know that there *may* be a problem before there *is* a problem.

One goal is to eliminate specific inspection points as everyone begins to own quality; a measurement focusing on the number of inspections may be useful in driving this number down. The focus in on the internal customer—the person who gets the product next—rather than on the external, or final, customer.

Traditionally, inspection departments are there to sort out problems. The emphasis is on not shipping bad-quality product; if a problem occurs, more inspectors can be added to catch defective merchandise before it is shipped. Of course, by this time, there is no way to clearly identify the cause of the problem or to solve it once and for all.

In the past, there has been very little emphasis on visible reporting of quality, especially by the people who make the product, on a daily basis. By implementing a "quality check as you go" by the people building the product, problems can be identified and solved quickly.

3. Using Cycle Time to Eliminate Waste

VARIANCE TO STANDARD → CYCLE TIME/FLOW

Cycle time refers to the elapsed time between the start and completion of a particular product. Focusing on this powerful measurement will have strong indicators associated with it: quality and work-in-process inventory. Remember, if cycle time is going down, you can be sure that quality is improving and that you have less WIP inventory than before. As the cycle time decreases, customer responsiveness increases. Implementation of flow and demand pull will drastically reduce the total cycle time of the product by eliminating nonvalue-added activities (see Chapters 6 and 7).

Standards are used primarily to cost products in a JIT environment, not to figure variances. A "stake" should be "driven into the ground" at the beginning to provide a constant measurement. One client has used this stake to measure actual cycle time against theoretical cycle time, or standard labor hours, for each product. The goal is a three-to-one ratio. In order for this type of measurement to work over time, the stake, or standard, cannot be changed; the measurement is against a constant.

Traditionally, standard costs for labor have been established and compared with the actual time it takes to perform a particular task; variances from standards are then determined for labor, material, and overhead. A positive variance indicates bad performance, while a negative variance indicates good performance.

In traditional environments, standards have also been used for detailed scheduling of operations; as material is tracked through each operation, actual data are collected in detail. A massive amount of data is compiled, and once or twice a year standards are updated according to them.

Let's take a closer look at how standards are used to schedule operations and how variances are calculated. The following process has three operations (see Figure 14-1).

Operation A has a standard time of two hours, operation B has a standard time of two hours, and operation C has a standard time of three hours, resulting in a total standard time of seven hours. Actual times show a zero variance for operation A, a positive (bad) .5 variance for operation B, and a negative (good) .5 variance for operation C.

The operations each had a scheduled time to begin according to standard. Since operation B was over by .5 hour, operation C was late getting started. Luckily, workers performing operation C hurried and made up the .5 hour, finishing on time.

During several years of asking companies what percentage of their standards are absolutely correct, I have usually received an answer that it is in the range of 10 to 20 percent. Many cost accountants are kept busy just figuring all the variances for each batch. In the end, some variances are over, and some are short. The questions to ask are:

- Do these detail measurements really uncover problems?
- Do they add value to the product?
- Does this measurement result in better business decisions?

If the answers are no, then eliminate them.

To implement cycle time measurement, measure actual time globally—cycle time for the line or product—and schedule by building what is

Figure 14-1. Using standard hours to schedule operations.

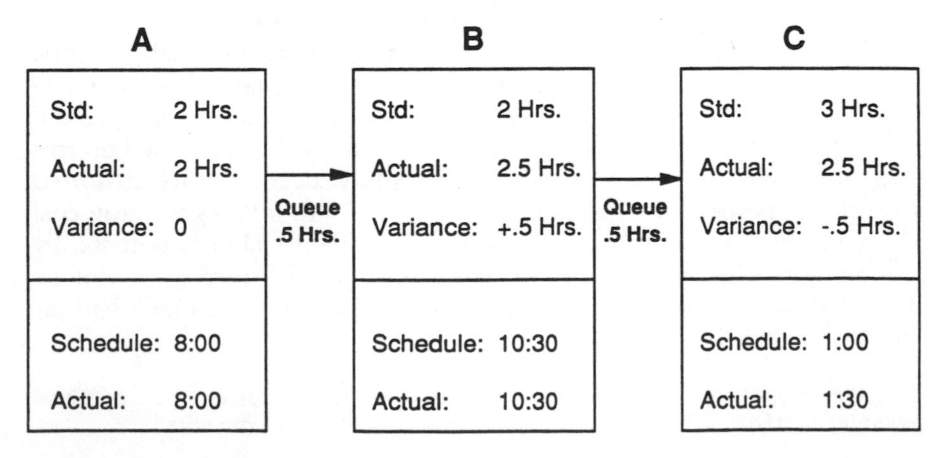

needed according to a daily regular schedule and visible demand signals. A focus on reducing total cycle time drives the waste out of the process.

In a 1988 *Harvard Business Review* article, George Stalk argued that as cycle times are reduced significantly by removing waste in the process (usually cycle times are reduced to 20 percent to 30 percent of the baseline when waste is removed at a reasonable level), productivity increases twofold and total cost of the product can be reduced by 20 percent.*

4. Making Labor Costs a Part of Overhead to Focus on Flow

LABOR DETAIL/OVERHEAD BASE → GLOBAL INFO/FIXED COST

Under JIT, labor, both direct and indirect, is a part of fixed cost, or overhead. You should emphasize attaining the Drumbeat every day with a smooth flow of product, not collecting detail labor or distinguishing between direct and indirect labor.

The right amount of labor capacity is present when Drumbeat is achieved. When the flow of product becomes more visible, capacity requirements become visible as well; there is no need for detailed labor reporting to determine how things are going. Detailed tracking of labor through work orders is eliminated, and overhead is based on throughput or cycle time or is allocated to each product line directly.

Traditionally, labor hours have been collected in detail per person per work order per operation. In Figure 14-1, detailed labor calculations are collected at each operation, and a variance against standard is computed. This requires that the person input the information at each step of the way. Real life says that most people input their labor at the end of the day and that they estimate times for each work order anyway.

Cost systems typically allocate overhead dollars to direct labor hours. As the direct labor content has gone down over the years, this has made less and less sense. For instance, labor typically contributes 3 to 5 percent of cost of goods sold in high-tech industries. All overhead is based on this 3 to 5 percent of total cost. Make-versus-buy decisions are distorted because the percent of overhead applied to all labor inflates the true cost of the product. If the fraction of costs attributable to labor is reduced by improving the process, a greater percentage of the total cost will be attributed to overhead. A total reduction in cost actually can look bad, as shown in Figure 14-2.

*For a more detailed discussion see "Time—The Next Source of Competitive Advantage," *Harvard Business Review*, July–August 1988, by George Stalk.

Figure 14-2. Effect of productivity improvements on traditional overhead measurement.

Product × Standard Costs: (Before Cost Reduction)		
	7.00	Material
	1.00	Labor
	2.00	Overhead
Overhead = 20%	$10.00	

Product × Standard Costs: (After Cost Reduction)		
	7.00	Material
	.30	Labor
	2.00	Overhead
Overhead = 22%	$9.30	

The ratio of direct labor to indirect labor is another traditional measurement associated with labor hours. With JIT, it makes no difference—in fact, you want to remove all barriers and ask your people to do what makes sense for success.

5. Measuring Assets to Ensure Effective Utilitzation

INVENTORY TURNS/ASSET → INVENTORY DAYS ON HAND/ LIABILITY

Asset ratios relate total sales to asset dollars. The idea here is to turn investment into sales as quickly as possible.

Traditionally, inventory has been seen as an asset, and the ratio has been stated in inventory turns. This calculation relates total inventory to sales—a global number that is very difficult to translate to a particular product or area and is therefore easy to ignore.

In a JIT environment, inventory measurements are personalized by using "days on hand." Most people can easily relate to the number of days' supply rather than lead time to buy or make more product for their own area. Too much inventory quickly becomes visible.

Employees are certainly another asset. Instead of focusing on labor hours, you may find it useful to focus on throughput per employee. Measure dollars spent training this very valuable asset—your people. Training is more essential than ever before as people take more responsibility and ownership for the products being produced.

Space is another very expensive asset. Some companies measure space per output. Others have put a stake in the ground and simply measure total space to the original baseline. This number can be distorted quite easily, however, if volumes change. When space is freed up, it should be clearly marked. Put a ribbon around the area and see it as space savings—make it visible.

One client had a major cleanup and reorganization as part of its housekeeping step. The company found a lot of unused junk in the process. People couldn't decide what to do with it, so they put it in its own area, roped it off, and labeled it, appropriately, "junk."

Traditionally, the only time companies worried about space was when they ran out of it. The immediate assumption was that another building was needed. Today, people have learned that in many cases space can be doubled just by making the changes outlined in Chapters 5, 6, 7, 8, and 9 on inside-the-factory issues.

6. Focusing on Supplier Partnership to Improve

PRICE/DELIVERY → PARTNERSHIP/QUALITY

Suppliers should be measured in a consistent manner to identify opportunities for improvement. These measurements are effective only when they are agreed upon by you and your supplier. In the past, suppliers have been measured first and foremost on price and second on delivery.

For JIT, measure the number of suppliers, quality, promptness of deliveries, batch sizes, and other issues discussed in Chapter 11.

7. Measuring the Setup Time Reduction to Increase Flexibility

TO STANDARD/EOQ → REDUCED TIME/MORE SETUPS

Setup time in the past has had its own standard time, separate from that of production time. The goal was to make or beat standard—nothing more.

Setup time is considered part of the run time standard in most JIT environments. Focus on reducing the setup time to allow for more setups and therefore more flexibility. Lower setup times allow smaller batches to be built more often at no additional cost, providing the opportunity to make what customers need when they need it.

8. Costing Extra Capacity to Encourage 100% Utilization

BURDEN EXCESS CAPACITY → SELL EXTRA CAPACITY

Capacity costing here deals with unused capacity. If some portion of your facility is not being used at the present time, how would overhead be applied if you were to consider producing a new product that would utilize that space?

Traditionally, a straightforward costing model—figuring out the direct-labor hours needed to build the new product and applying the plant overhead to those labor hours—would be used to see whether there was capacity.

JIT says to apply common sense to the things you do—obviously it would be beneficial to put this unused space to good use. If considering a new product, then, apply the variable overhead dollars only when calculating the cost for the new product—the space is there whether you use it or not. In other words, sell your extra capacity.

9. Justifying Capital Expenditures to Support Flow, Balance, and Quality

RETURN ON INVESTMENT → SUPPORTS JIT AND TOTAL COST

Capital expenditures, as well as make-versus-buy decisions, are viewed differently in a JIT environment. Consider the support of the JIT vision for the company above and beyond the direct costs associated with the investment.

Equipment may be selected and justified to support flow, balance, or quality of the product. Simpler equipment that is more flexible may be chosen over a more expensive does-everything piece of equipment.

Most return-on-investment models do not take into account such things as the cost of quality. Change ROI calculations to show total cost, not just direct cost.

As mentioned earlier in this chapter in the discussion on overhead allocation, make-versus-buy decisions can be distorted by high overhead numbers caused by the decrease in labor cost as a percentage of the cost of goods sold. Other considerations are also important in a make-versus-buy decision.

Measure the Right Things

Recently a company was evaluating the possibility of making parts in-house rather than continuing to purchase them from a company in the Far East. The firm was experiencing some of the typical problems of having parts made outside, including a huge pipeline of parts, which resulted in little flexibility for change and limited the company's ability to react if something were to go wrong. In addition, the purchaser had very little control over how the parts were made or their quality.

Despite the cost of these problems in terms of customer service, the accountants were willing to look only at direct costs, which had a large overhead burden associated with them. The decision was to continue to buy the parts overseas. Everyone knew that the total cost was higher with the overseas parts, but the cost accounting measurements forced the wrong decision to be made.

10. Focusing on Long-Term Success to Get Problems Solved

"SHIP IT"/EARNINGS PRESSURE → LONG-TERM FOCUS

With JIT, focus is on long-term success with continuous improvement. This may mean that short-term results will suffer to make things better in the long term. This idea is one of the most difficult to execute when the going gets tough as in that age-old dilemma: quality versus "ship it."

With JIT, watch short-term trends carefully and pay special attention to changes, but focus on longer-term success indicators.

In traditional settings, every focus is on the shipment dollars, which causes a "ship-it" attitude to prevail. Short-term earnings pressure is so great in some companies that it is impossible to take the time to focus on and solve some of the long-term problems.

11. Encouraging Innovation to Drive Improvement

THE WAY WE ALWAYS DID IT → FOCUS ON NEW IDEAS

In the past, innovation, or the implementation of new ideas, has taken a back seat to shipping product. However, it is the essence of continuous improvement. Therefore, you should include it as a measurement to drive the change process.

Innovation may be measured as success in responding to customer needs, the number of new ideas or suggestions implemented, or the number of new designs created.

12. Measuring Total Customer Service to Increased Market Share

SHIPMENT GOALS → PARTNERSHIP/INCREASED MARKET SHARE

Customer service really is the ultimate success measurement. Success with many of the measurements mentioned in this chapter is required for the ultimate in customer service. To be successful in today's marketplace, you must focus on total customer service instead of on many of the micro measurements of the past. A partnership—"both-win"—relationship should be established with the customer. Excellent customer service can increase market share.

Costing Methods to Drive the Change to JIT

It is clear that traditional costing systems are not adequate for a JIT environment; it is much less clear what the ideal costing system is. However, some themes seem to be emerging as companies get more experience with JIT. These include:

- Basing overhead on throughput or process time rather than on a labor-hour base.

- Basing material overhead on material cost and the rest of overhead on something else.
- Treating each product line like a separate business with its own profit and loss. Such product-line costing creates ownership, and sound decisions result.

Product-line costing requires that all costs be allocated to the particular product line as they are incurred. This method makes costs much more visible for the people associated with each line, leading to more effective cost management.

Some costs, such as materials and labor, are very easy to tie to a particular product line. Even building usage can be allocated on a square-foot basis for manufacturing. But what about all those indirect costs, support services, and management? These are much more difficult to allocate and require detailed systems to do so.

Some companies have gone this route just to get a handle on which products are really generating profits. One client did this and found that out of five product lines, three were making money and the other two were losing it all!

In order to measure the savings that result from product-line costing, it is necessary to first baseline where you are today. Allocating the indirect costs is a painstaking but necessary effort. As you begin this process, one product line at a time, the accountants will claim that all you are doing is pushing the indirect costs to other products. But the client who went through this process believes that once you get about 30 percent of the plant on product-line costing, you begin to realize real savings in indirect costs. So the message is: This is not a short-term process—it takes time and practice to find and realize the savings.

Rather than creating a detailed system to allocate the difficult-to-identify indirect cost, why not allocate what you can without creating another detail monster and allocate the remainder to cycle time? This idea has been implemented by Gary Flack at Zytec. Cycle time is something that you always want to drive down, so by allocating indirect costs, or overhead, to cycle time, you create the incentive for shorter cycle times and allocate less overhead. Figure 14-3 illustrates this concept.

In this example, there are two product lines: A and B. Line A has a cycle time of one week, and Line B has a cycle time of two weeks. Any overhead not easily tied to a particular product line is allocated by cycle time. Since Line B has twice the cycle time of Line A, Line B is allocated two thirds of the total overhead dollars. Surely Line B will be working to reduce cycle time to avoid this overhead expense.

You may be wondering about how to time the change in measurements and, particularly, costing methods for your products. I have found that

Figure 14-3. A costing method that drives a reduction of cycle times.

PRODUCT LINE COSTING

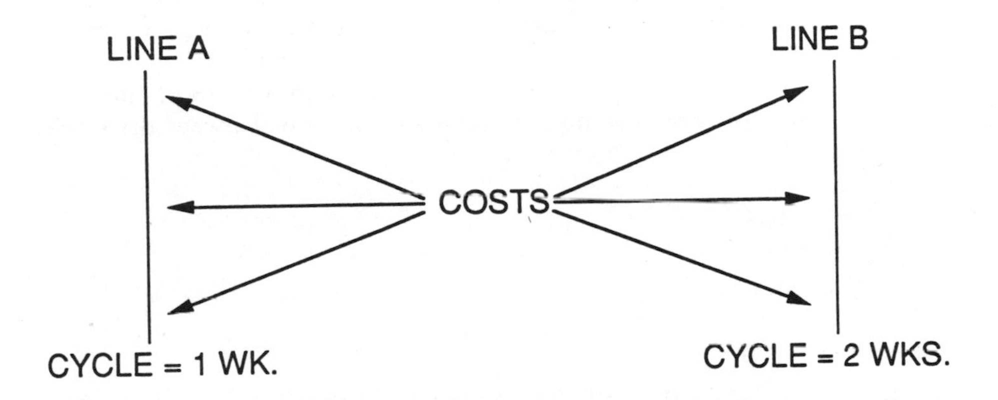

Overhead Difficult To Allocate Based/Allocated On Cycle Time

CYCLE TIME ALLOCATION OF OH:

1/3 = LINE A
2/3 = LINE B

P&L BY PRODUCT LINE

these changes follow some initial experience with JIT in manufacturing and/or nonmanufacturing functions. Measurement and costing systems seem to change only when the existing ones become a barrier to further improvement.

Companies that begin to worry about these changes before they are a barrier will speed the process of continuous improvement.

Three Measurements to Ensure Success With JIT

It is vital that you pick the right few performance measurements—those that can be key indicators for the health and growth of your company. Everyone throughout the organization should understand clearly what these measurements are, why they are important, and what they can do to help. The number of measurements needs to be small enough so that everyone can name them at a moment's notice—a maximum of three or four. The three performance measurements I believe will always drive the change are:

1. Drumbeat
2. Quality
3. Cycle time

Performance Measurements for People to Support JIT Principles

Performance measurements for people should drive the cultural change to JIT. It is necessary to pay attention to this issue early in an implementation; temporary performance measurements can be put in place even in an early pilot for the people involved. Human resources personnel should be involved in JIT from the beginning, not only so that they can help in the initial stages but also so they can learn to allow this all-important function to support the changes long-term.

A Lesson Learned!

In one company, it didn't take long for "building only when it's needed" to make sense. The norm in the pilot area became building the right parts at the right time, when they were needed. No longer was this line building as many as possible. The people were working as a team, improving the process. Quality and customer service improved.

People felt good about their success until performance evaluations were done. The evaluation criteria had not changed—the form was the same as always. Quantity and efficiencies were high on the list of performance criteria. The people became very confused: They were being told one thing on the line and evaluated on something else. Management quickly realized that it had forgotten a critical piece of this change process—the measurements. A temporary evaluation form was developed with the help of human resources for all JIT pilot participants to correct this oversight.

The traditional measurements for people in manufacturing emphasize:

- Quantity
- Working harder
- Working fast
- Specialization
- Sorting to identify problems

New performance measurements emphasize:

- Teamwork
- Working smarter
- Process improvement
- Flexibility
- Ownership for quality

Consistency between actions and performance evaluation criteria are necessary and critical as you embark on this change process. Several client companies have changed the evaluation criteria for the people involved in JIT pilots by using a new evaluation form or an addendum to the present form.

Form Follows Function

AGFA Compugraphic had a unique approach to developing the new evaluation form—one that supports the JIT mindset completely. At AGFA Compugraphic, the human resources manager had been involved with the JIT implementation since the very first pilot.

Within a short time after beginning the first JIT pilot, this manager was getting complaints that "the old appraisal is contradictory to what we are being told about JIT." The sense was that, as at HP when I was there, people were being graded individually and told to work in teams and that they were being evaluated in terms of quantity when they were being told to stress quality, Drumbeat, and line balancing.

The manager brought the issue to top human resources managers and even took them on a tour of another company's facility that worked in a JIT environment. The top managers at AGFA Compugraphic thought that JIT was just another manufacturing thing that would go away. But it didn't, so they told the manager to work on a new evaluation form.

This manager found by surveying other companies implementing JIT that in few, if any, cases had the human resources department been involved in the effort. With no guidance from other companies, he decided to go right to the experts—the people on the pilot teams—to ask them what they thought was important about the work they did and on what basis they should be evaluated.

The old evaluation form had been developed by middle and upper management, and the workers had never been involved in such a task.

The manager chose two teams: a first-shift team that was having some difficulties working as a team, and a second-shift team that had complained that it was being left out of extracurricular activities because it was on the second shift. He asked each team to develop an ideal evaluation form within four to six weeks.

He made no guarantees about accepting their proposals—only that he would not change what they had developed without asking them.

The teams went through several stages:

1. Disbelief. They didn't believe that management was really going to let them do this.
2. Anger. They found this to be the most difficult task they had ever undertaken.
3. Involvement. They finally realized that the manager was serious and began to develop what they thought was the right evaluation form.
4. Completion by both teams within six weeks.

The forms were very similar. The manager combined them, got approval from both teams, and took the form to top management. The form was approved as it stood. Exhibit 14-1 at the end of the chapter shows the old and the new evaluation forms. Since then, the new form has been replaced by a universal corporate form.

Revamping the evaluation form is only a first step. Another client company asks each employee to do a self-evaluation and gives people the opportunity to discuss with the manager his or her evaluation of them. At another company, evaluations are done in a customer/supplier fashion, and each person may be evaluated by two dozen or more internal "customers."

Effective Reward Systems for JIT

Today's management has a major concern about reward systems for people involved in JIT implementation. The concern goes something like this: "If I ask my people to get involved and contribute at a higher level than before, they will expect something in return." Of course, we immediately think of money as the "something in return."

There are two problems with this thinking. One is that many times companies implement JIT in order to compete more effectively; therefore, there is not an excess of funds to be passed around. Second, money seems to provide only a spurt of motivation.

Suggestion systems have been implemented in many forms for years—long before JIT. We had several at HP. Each involved formal paperwork, reviews, and winners. In each case, the number of suggestions went down over time, and no one could figure out why.

It has now dawned on managers why these never work—there were a few winners and many losers. Who wants to keep offering suggestions after repeatedly being a loser?

Guard against any program in your company that has winners and losers. The goal is to have all the suggestions be winners. By creating empowered teams that can implement their own ideas, you can ensure that the good suggestions get implemented.

Another option that has been popular in some companies is gainsharing—a program that allows employees to share the benefits of improvements as they are made by groups. This is an excellent option for those companies that have individual incentive pay—that pay employees for each piece produced. Of course, individual incentives are barriers that must be removed before implementing JIT, since people will no longer be allowed to build as many as possible. You would be asking employees to accept a cut in their income if they were to follow the new rules of building only when it is needed.

My experience has been that most people like being more involved through teams. They feel that being empowered to solve some of their problems makes for a better place to work. Some very effective rewards are a visit by top management to congratulate workers or to inquire about their concerns, a letter from management letting workers know how they have contributed to the success of the company; pictures and descriptions of accomplishments displayed in visible places, such as on the company bulletin board; plant meetings in which the teams present accomplishments; or small celebrations in the form of cookouts, pizza, or an extra break with refreshments.

One company created new break areas for employees; another improved the cafeteria—with money generated from the success teams had in reducing production costs.

The most effective recognition is a regular, consistent acknowledgment of a job well done.

Quantitative performance measurements are helpful in driving the change process associated with JIT. However, qualitative measurements—those more difficult to measure—should be watched as well. They include:

- Understanding of JIT
- Education

- Change in management perspective
- Management follow-through
- Quality of work life
- Customer satisfaction
- Capacity to respond to opportunity
- Ability to master technology

Very carefully designed surveys have helped in giving managers a better handle on the people issues and the customer satisfaction issues. However, there is really no substitute for having a close relationship with employees and taking the time to talk to them.

As the company changes to JIT, performance measurements—both financial and people—can drive the change. Some of the keys for success including making these measurements simple, visible, less detailed, fewer in number, and value-added only.

Exhibit 14-1.(a) Old form.

Old Form

cg compu**graphic**

Review Period ____________

Performance Appraisal

Employee________________________ Dept.__________ Position________________

RATE ON FACTORS BELOW	UNSATISFACTORY	FAIR	GOOD	VERY GOOD	EXCEPTIONAL
1. QUALITY OF WORK. Extent to which work produced conforms to the requirements of the job.	☐	☐	☐	☐	☐
2. QUANTITY OF WORK. Volume of work regularly produced. Speed and consistency of output.	☐	☐	☐	☐	☐
3. DEPENDABILITY. Extent to which employee can be counted on to carry out instructions and fulfill responsibilities.	☐	☐	☐	☐	☐
4. JOB ATTITUDE. Amount of interest and enthusiasm shown in work.	☐	☐	☐	☐	☐
5. ATTENDANCE. # of days absent ______ # of occurrences ______ # of Mon/Fri singles ______ # of days other (e.g., LOA) ______ **Comments:**	☐	☐	☐	☐	☐
6. PUNCTUALITY. # of days late ______ # of days left early ______ # of days other ______ **Comments:**	☐	☐	☐	☐	☐
7. SAFETY ATTITUDE. Measure of employee's concern for safety and compliance with safety standards.	☐	☐	☐	☐	☐

8. HOW DO YOU EVALUATE THE EMPLOYEE'S OVERALL PERFORMANCE? LIST BOTH STRONG POINTS AND WEAK POINTS.
Comments: ____

Exhibit 14-1.(b) Form revised for JIT.

New—Developed for JIT

cg compu**graphic**

Performance Appraisal

Review Period ____________

Employee______________________________ Dept.___________ Position____________________

RATE ON FACTORS BELOW	UNSATISFACTORY	FAIR	GOOD	VERY GOOD	EXCEPTIONAL
1. QUALITY (same as original)	☐	☐	☐	☐	☐
2. TEAMWORK (addition to original incorporates Job Attitude - see attached)	☐	☐	☐	☐	☐
3. PRODUCTIVITY (replaces Quantity)	☐	☐	☐	☐	☐
4. DEPENDABILITY (same as original)	☐	☐	☐	☐	☐
5. EVALUATE OVERALL PERFORMANCE LEVEL (new—summary similar to salaried)	☐	☐	☐	☐	☐
6. ATTENDANCE. days absent ____________ occurrences ____________ Mon/Fri singles ____________ days other (e.g., LOA) ________ **Comments:** (same as original)	☐	☐	☐	☐	☐
7. PUNCTUALITY days late ____________ days left early ____________ days other ____________ **Comments:** (same as original)	☐	☐	☐	☐	☐

8. HOW DO YOU EVALUATE THE EMPLOYEE'S OVERALL PERFORMANCE? LIST BOTH STRONG POINTS AND WEAK POINTS. INCLUDE CONCERN FOR SAFETY.
(safety changed to essay response—delete blocks)

____

____

____

____

____

8. (continue, if required)

9. WHAT IS EMPLOYEE DOING AT PRESENT TIME TO IMPROVE PERFORMANCE? WHAT ARE YOUR SUGGESTIONS FOR HELPING THE EMPLOYEE PROGRESS IN PRESENT JOB? FUTURE JOBS?

10. WHAT ARE THE EMPLOYEE'S PERFORMANCE OBJECTIVES FOR THE NEXT REVIEW PERIOD? (At least one objective should address quality of work.)

WRITTEN BY ______ DATE WRITTEN ______

SIGNATURE OF REVIEWER ______ DATE APPROVED ______

SIGNATURE OF REVIEWED EMPLOYEE ______ DATE DISCUSSED ______

TEAMWORK: Amount of interest and enthusiasm shown in working with others toward completion of common goals.

UNSATISFACTORY: Unable to work with others. Attitude too poor to retain in job without improvement.

FAIR: Not fully participating and needs improvement to be acceptable.

GOOD: Able to work effectively with others at an acceptable level.

VERY GOOD: Above average ability to work with others with a high degree of enthusiasm and interest.

EXCEPTIONAL: Outstanding ability to work with others, generate ideas, resolve conflict, and solve problems.

PRODUCTIVITY: Ability to meet work flow requirements.

UNSATISFACTORY: Output is inadequate to retain in job without improvement.

FAIR: Output is below work flow requirements.

GOOD: Output usually satisfies work flow requirements.

VERY GOOD: Output consistently satisfies work flow requirements.

EXCEPTIONAL: Output consistently satisfies work flow requirements. Able to readily adapt to emergencies, unusual demands, and/or new requirements.

Chapter 15

The Road Map for Implementing JIT

There are many ways to begin a JIT implementation. The particular option you choose must make sense for your company. There is no "cookbook" for implementing JIT—each corporate culture, management approach, marketplace, product, and process is different, and each company will begin from a different stage, with different issues. However, this book is about an approach that works—use it as a road map, but use common sense as you make the journey.

The following discussion is based on the assumption that your company has only one entity, or facility. If your company has a corporate function tying many entities together, the *vision* and *strategy* steps plus some of the *organization* step may take place within a corporate team if this is to be a corporate effort, or within the leadership structure of a particular division, business unit, or plant.

The leader of the company plays a key role in the implementation of JIT. Ideally, the first step is to get top management commitment to implement JIT concepts and techniques. Some companies have started in the middle—midway in the organization—and later obtained top management's attention through some successful results with JIT. However, this approach is very dangerous; you run the risk of getting people at the lower levels of the organization excited and energized to make changes and then running into a barrier if top management is not convinced that JIT is the way to go. It is better to take more time with top management up front and to get the leader's commitment, this ensures that the change process will be successful.

To ensure top management commitment, you can provide JIT awareness sessions, arrange visits to other JIT companies, offer JIT books and articles, and develop cost/benefit models for JIT within your company.

A Success and a Failure

In one recent JIT implementation effort, we started in the middle and it worked. The culture encouraged people to try new things when it made sense, so we did. After some successes, top management began to notice and eventually became committed to this new way of doing things.

Early in my consulting career, I thought that this approach should always work—after all, how could management argue with success? How could it not get excited and committed?

We began with a certain client in the middle. One plant manager in a division that is a part of a larger corporation wanted to implement JIT. The divisional management was willing to let the plant give it a try; corporate, which included product engineering and marketing, was oblivious to the whole effort.

The plant was very successful, and, eventually, that success spread to other plants in the division. The thrust continued to come from the plant management, however.

Consistency between the plants was lacking, because the effort was not being driven from above.

However, these manufacturing facilities were very successful in reducing cycle times, achieving a Drumbeat, and increasing quality. Quality of work life had improved significantly as the people felt empowered to solve problems.

At that point, issues became product design, measurements, purchasing, and marketing. Corporate was still not involved, and manufacturing had no power over these important functions that now could block future progress.

Some typical questions about getting started include:

- How do we begin?
- What are the issues that we should worry about?
- Who should be involved?
- When are we ready?
- Where should we begin?
- What are the benefits?
- What is the cost/payback?

The implementation itself is carried out in six phases, the timing of which is shown in Figure 15-1.

In the sections that follow, I discuss implementation steps and the results you may expect within each of these phases. This is followed by a discussion of benefits to expect with JIT.

Figure 15-1. Timing the six phases of the JIT implementation process.

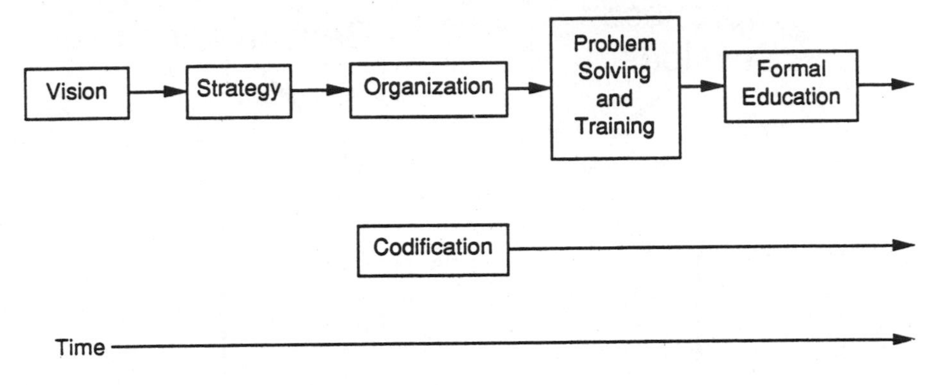

Creating a Vision to Get Commitment

The vision of "what could be" is created by top management (see Figure 15-2). It is necessary for these leaders to understand the principles of JIT and how they apply to your company, as well as to determine the opportunities and issues your company faces. Awareness sessions can be held to introduce top management to the principles of JIT. Once the major issues facing your company are determined, it is wise to investigate those companies that do particularly well at solving these issues—a process often called benchmarking.

Awareness of what is possible and of what questions need to be asked should emerge from this step. Enough dissatisfaction with the present mode of operation should be generated to get top management commitment to change.

Determining Opportunities and Needs to Create an Improvement Plan

During the strategy phase, opportunities and needs are assessed (see Figure 15-3). The vision of how your company could function becomes clearer, and a plan for improvement is formulated.

The opportunities for your company can be identified in a global sense through a study that includes plant tours, interviews, and data gathering about current conditions. A more detailed look at what could be can be achieved through an Opportunity and Climate Assessment (OCA). An OCA can help determine the specific, realistic, and measurable opportunities available through JIT, whether the company's organizational climate is supportive of the needed changes, and whether the company is ready technically to make the change.

Figure 15-2. Creating the vision.

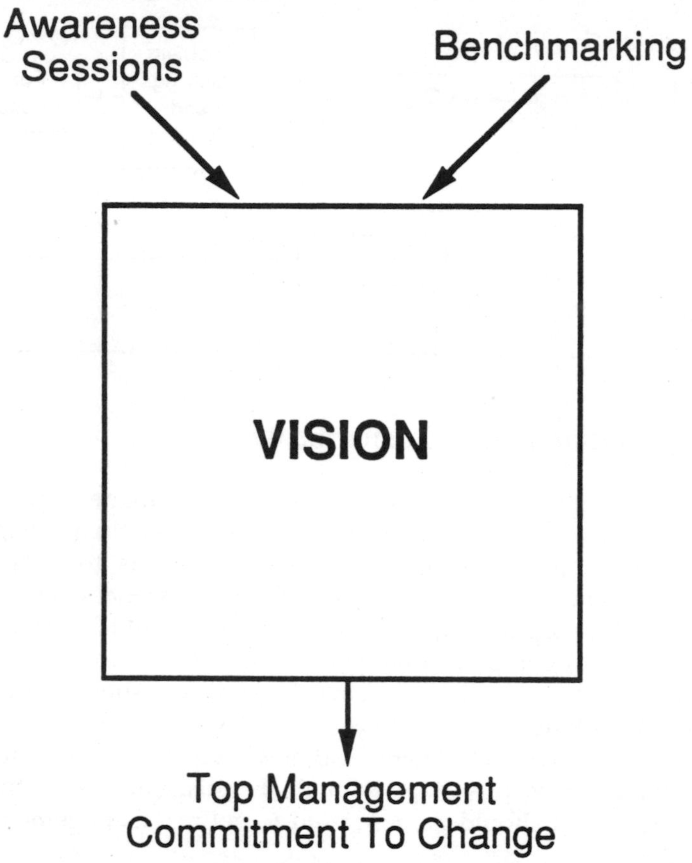

A Customer Needs Survey may also be appropriate to determine what is important to the customer. Understanding specific customer needs can be most effective in driving your company to change.

An Era Management Plan—a top-management plan to manage a discrete time frame (usually between 18 and 36 months)—can also be developed by the leader. This is the leader's plan for improvement, both on a personal and corporate level; the leader often changes personal behavior in order to create a model for other company personnel to follow.

By the end of the strategy phase, top management is convinced that change is in order and has developed a vision of how things could be, as well as a plan to get there.

Organizing for Results

A steering committee should be formed to guide the implementation effort (see Figure 15-4). This committee is usually made up of the leader and the

Figure 15-3. Determining opportunities and needs.

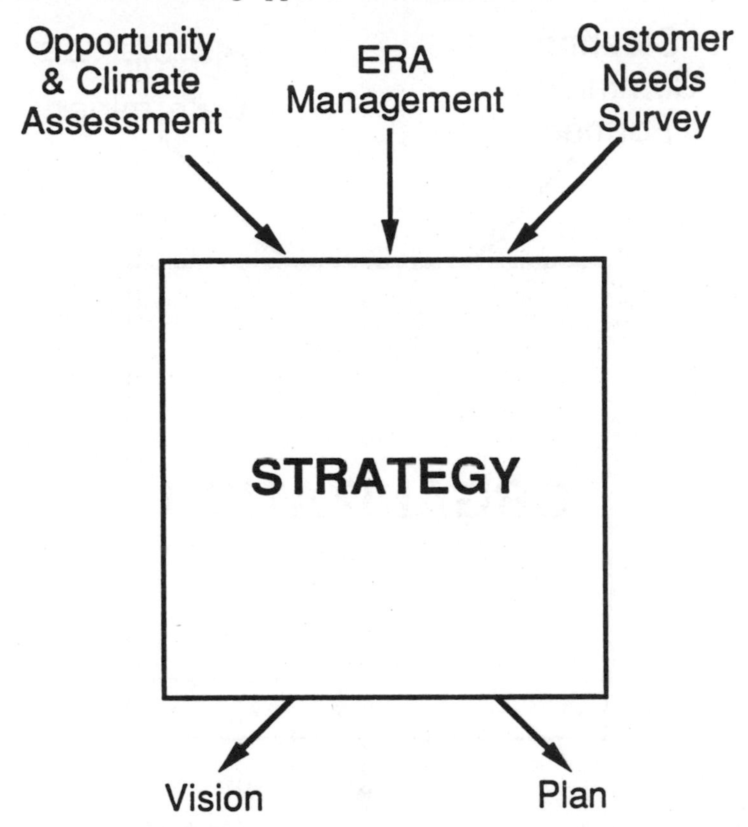

top management group; it is important that all corporate functions be represented on this committee, even if they are not presently part of the management committee. Another alternative that has been successful in some companies is a cross-functional and cross-levels steering committee. The risk with this type of committee composition is that some level of management may be "left out of the loop." With any team, the ideal size is between five and ten members. A coordinator for the implementation is also required. This could be a full-time job in a larger organization or a part-time responsibility in a smaller one.

Starting Small With Pilots

A JIT implementation should begin with pilots. Pilots allow you to start small and build on your successes and give you the opportunity to discover how JIT works in your company and to experience its benefits without great risk.

Figure 15-4. Organizing for implementation.

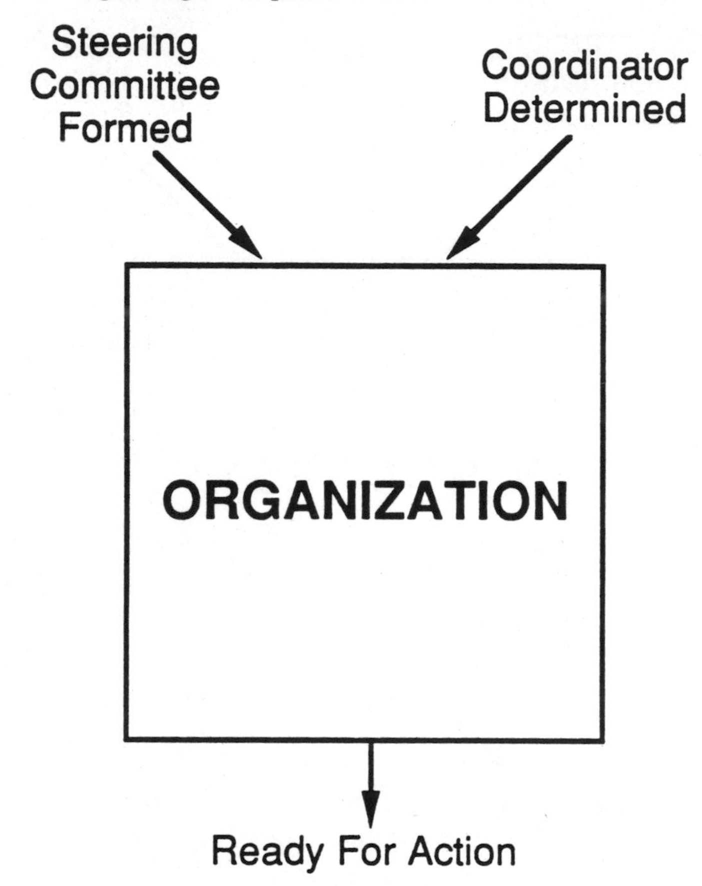

If an entire company implementation were done, there would not be adequate resources to support all the teams and ensure their success; the entire effort would fail miserably. Pilots, on the other hand, allow you to plan the implementation one step at a time and, as you acquire more experience and knowledge, to develop an implementation approach that works for you. It is important to keep a diary or journal to document this approach for the future.

Determining the Steering Committee and Its Charter

The steering committee will manage the implementation of pilots. Tasks included in the steering committee's charter include the following:

- Picking the pilots and pilot teams
- Determining the scope and time frame of each pilot

- Developing an implementation plan for each pilot
- Developing a mission statement for each pilot
- Further developing the company vision through pilot experience
- Ensuring that learning from pilots is shared
- Establishing key measurements for each pilot
- Monitoring results of pilots through regular reporting
- Dealing with barriers
- Providing support and resources for pilots
- Monitoring and approving changes when necessary

Now it is time for some action.

Developing an Implementation Plan That Results in Successful Pilots

In order for the steering committee to fulfill its charter, it will need to develop a very good and detailed understanding of JIT principles and techniques, plus effective skills in working as a team. Education in both JIT and effective teamwork is the next step. As part of the team-development process outlined in Chapter 3, the team can then develop its goals and ground rules. Now the steering committee is ready to pick the pilot (see Figure 15-5).

Figure 15-5. Developing an implementation plan.

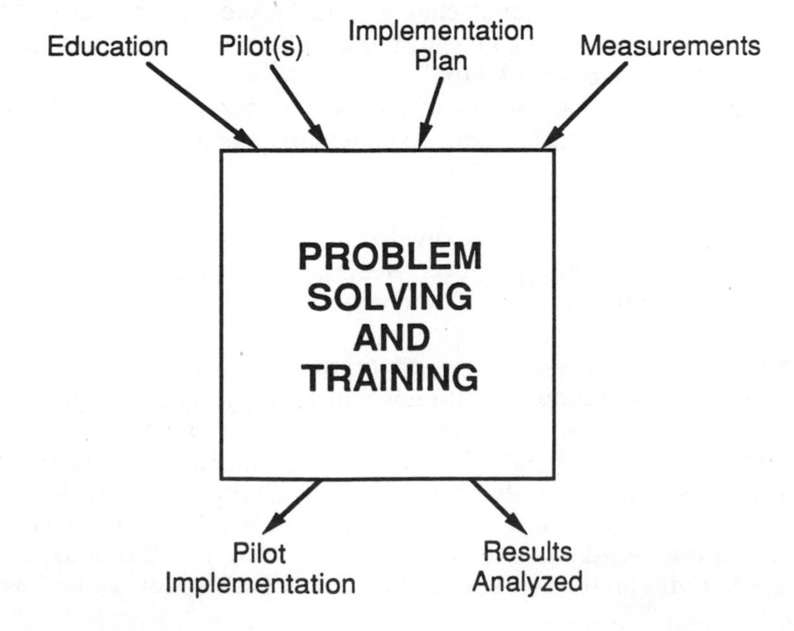

Picking a Pilot

One way for a steering committee to begin its work is to make a list of all the possible pilots—administrative and indirect areas, as well as manufacturing. Then make a list of the criteria that are most important for your company. Finally, rate each idea against the criteria to make the best decision.

A Starter Criteria List for Selecting a Pilot

Pick an area that is *small* in scope and number of people. Each pilot should be self-contained; a rule of thumb that I often use is not to involve more than twenty-five people in the initial pilot.

Pilot should be *visible*. One purpose of pilots is to sell the idea of JIT, so pick an area that can advertise the benefits.

Each pilot should be *representative* of the business. If others in the company can't relate the successes of a pilot to their areas, you get limited benefit.

Setting an Example

We began our efforts in the printed-circuit-board fabrication area in Boise. The success was high—suppliers were delivering within four-hour windows every week, quality was up, customer service had improved. However, when people in the disc-drive manufacturing area looked at our success, they replied, "Oh, sure, it will work for you! You only have ten direct materials! We have 3000! It will never work for us."

Eventually, our team was move to a line making a new disk-drive product. We were successful there, and then the other disc-drive lines began to pay attention.

At Valleylab, the business is divided into two very different products: hardware and disposable products. Managemnt chose to do two pilots, one in each area, to show that JIT techniques would work throughout the company.

An important criterion for selection is that *process must be well understood,* with no major problems. If people are wrapped up in fighting fires or trying to understand their process, they will not be able to devote the necessary attention to a pilot. If a division or line has a problem that is well understood, however, that may be a very good place for a pilot.

Most important is a *high likelihood of success.* JIT pilots must be successful. You cannot afford a failure in the early stages of JIT. Take into account the people in the areas that you are considering for pilots, as well as the product and process. People who thrive on change and like the opportunity to be innovative and lead an effort are best for pilots.

Picking the Pilot Team

Picking the pilot team is usually straightforward once the pilot is selected. Make sure that everyone with a stake in the pilot is included. For instance, if you pick a manufacturing area, include everyone who works in that area, plus the manufacturing engineering personnel and possibly the planners who support the area as well. If you pick the product design cycle, make sure to include product engineering, manufacturing, manufacturing engineering, quality, purchasing, and marketing. These support functions can either be on call for the pilot team or full-time team members.

Defining the Pilot Scope

Once the pilot and the team are picked, the steering committee must clearly define the scope of the pilot. For instance, for a manufacturing area, the pilot may cover "the time the parts are received on the line until we move the finished product to finished goods." There must be definite walls on either side of the process involved in the pilot in order to measure effectively results from the changes.

At some point during the pilot, you will need a major checkpoint to determine how you are progressing toward your goals and to plan future steps. A common time for this checkpoint is four to six months into the pilot. The idea is not that the pilot is over at this time—it will surely continue and expand. But it is necessary to set checkpoints so that achievements can be evaluated. For instance, if the goal is to reduce cycle time by 50 percent, that may be a reasonable goal to achieve in six months, but a totally unreasonable goal for one month.

Writing the Implementation Plan

The implementation plan includes the mission statement, or the overall purpose that this project will accomplish for the company—the organization, the pilot and scope, the time frame—plus major goals, measurements, and milestones.

To set a goal:

1. Identify the results you want.
2. State evidence that will show when the result has been accomplished.
3. Ensure that the evidence is measurable.
4. Indicate a realistic time frame.

Goal setting at the steering-committee level should be global, but specific, leaving the pilot team an opportunity to set plans on how to reach the goals. Goals should force the teams to present new paradigms. For instance, a 10 percent reduction in cycle time can probably be achieved by working a little harder while doing things pretty much the same way. However, achieving a 50 to 80 percent reduction in cycle time will require a change in paradigms.

Once a paradigm-changing goal is set and adopted, most teams can develop a plan to attain the goal. For instance, one team may go after setup times first, and another may change the layout and the flow. With the straightforward problem-solving process and tools discussed throughout this book, a team can gather data on where time is lost in the process and begin to develop soulutions for reducing that time loss. Anytime a goal is set, however, it must be baselined, that is, compared to where the process is today. The way in which each goal is measured should be clearly stated. Exhibit 15-1 at the end of the chapter shows excerpts from one of Valleylab's first pilot plans.

Once this plan is completed, it should be communicated to everyone. Major milestones such as team kickoffs or special training events are set by the steering committee. A plant awareness session should be held to explain the what, why, and who of the JIT implementation. Then the pilot team can be approached and formed.

Again, the first step is in implementing the pilot education about JIT and effective teamwork. Refer back to Chapter 3 for details of this process.

Once the steering committee has set goals for the pilot, the pilot team can establish its own implementation plan in order to accomplish those goals. It is very important that the pilot team be involved in determining how it will achieve the goals set out for it within the boundaries set by the steering committee and how it will measure results against those goals. The involvement = ownership = commitment principle makes it so.

Exhibit 15-2 includes later plans written by pilot manufacturing teams from Valleylab, and Exhibit 15-3 includes later plans written by pilot indirect teams from Valleylab. Both appear at the end of the chapter.

Implement and Expand

Now it is time to implement the plans and to measure your results. Reporting results regularly is essential in this process; without reporting, it is difficult if not impossible to determine the effects of the pilot effort. Along with some of the key measurements outlined in Chapter 14, don't forget to take photographs and videos as you go along.

At the checkpoint time for the first pilot, the steering committee can analyze the results and make a decision regarding future steps. Typically, the results are very good, and the steering committee decides to expand the effort and start new pilots. At some point, the committee will feel it has gained enough experience to determine the direction for the company and to develop a plan on how to get there.

Ensuring a Continuous JIT Process Through Education

During the formal education phase (see Figure 15-6), you and your staff will develop enough internal education and training skills to make this effort never-ending without having to continue paying fees to outside consultants or trainers. Eventually, all employees will have been trained and will be involved in various team efforts.

Figure 15-6. Education to ensure continuous JIT.

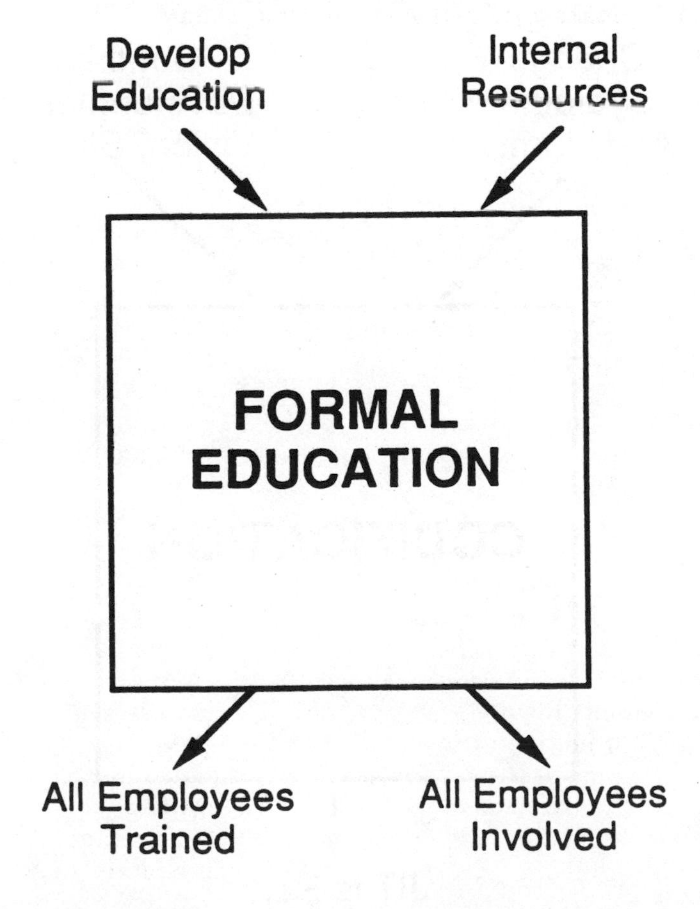

Making JIT a Part of the Business

Codification (see Figure 15-7) is the process of making the JIT mindset part of an organization's culture by changing the systems and norms people live by. This phase begins during the organization phase and continues indefinitely. Strategy and vision are developed further through a "second round" during this process.

The Test for Cultural Change

One way to determine if the changes you are making are permanent is to ask yourself what would happen if you went into your company and told everyone to stop JIT and go back to the way the company used to run. When you reach the point at which people would not listen to you and couldn't go back no matter what top management said, you will have made the change complete.

Figure 15-7. Making JIT part of the corporate culture.

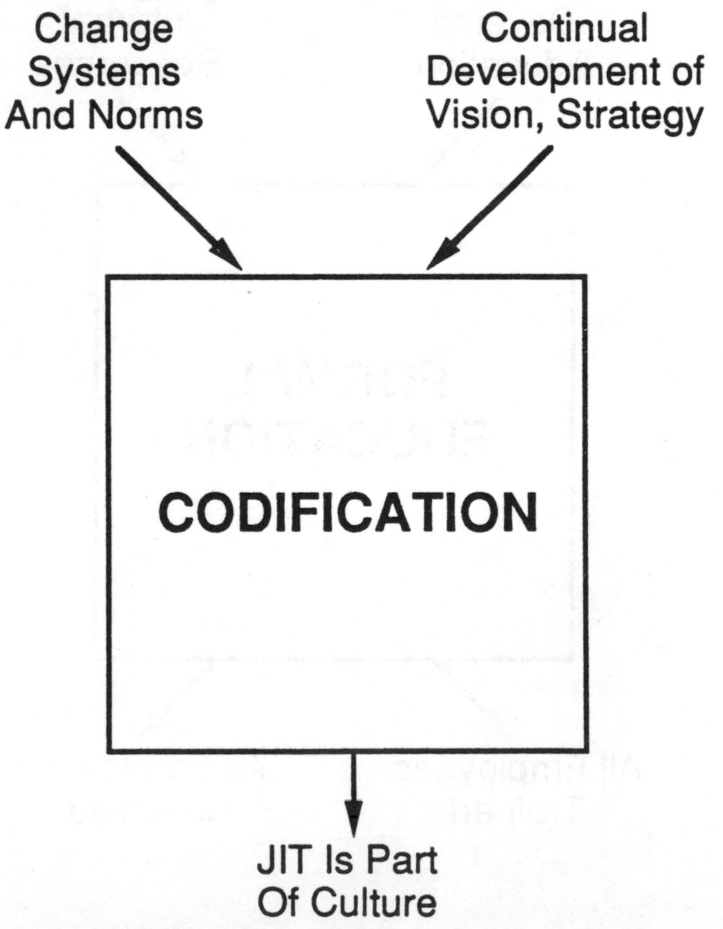

The Benefits Achieved With JIT

Throughout this book, I have shared some of the experiences of AGFA Compugraphic and Valleylab as they worked on implementing JIT. Figures 15-8 and 15-9 describe some of the benefits they have derived from the JIT philosophy and techniques. Typical results include:

- Cycle time: 80 percent to 92 percent reduction
- Direct labor productivity: 20 percent to 50 percent increase
- Indirect labor productivity: 20 percent to 60 percent increase
- Scrap and rework: 25 percent to 65 percent decrease
- Overall quality: 100 percent increase
- Purchased material price: 10 percent to 50 percent decrease
- Inventory:
  Raw material: 35 percent to 75 percent decrease
  Work-in-process: 70 percent to 90 percent decrease
  Finished goods: 60 percent to 90 percent decrease
- Setup time: 75 percent to 95 percent decrease
- Space needed: 40 percent to 80 percent decrease
- Capacity: 35 percent to 50 percent increase
- Customer deliveries: 100 percent on time

But even more important than these quantitive benefits are the qualitative benefits, such as:

- Improved quality of work life
- More rewarding and satisfying jobs for people
- Increased customer satisfaction

The bottom-line benefit could be survival for your company!

The Challenge to You

The challenge is change—continuous, consistent, and never-ending. The only way to achieve long-lasting success with Just-In-Time is through your people—all of them.

The approach outlined in this book works. Accept this challenge and make a difference. Your company can thrive in this competitive marketplace.

Good luck, as you

Make the Techniques
Really Work
by paying attention to the
Human Side of JIT

Figure 15-8. Improvements at AGFA Compugraphic resulting from JIT.

AGFA COMPUGRAPHIC

IMPROVEMENTS IMPACTED BY JUST-IN-TIME—1988

QUALITY IMPROVEMENTS

- Improved P.C. Blank Shipment Yield from 91.2 percent in Q1 to 94.8 percent in Q4.
- P.C. Assembly process maintained within S.P.C. control limits 9 out of 10 areas.
- Reduced P.C. Board Assembly/Test Cost.
- Reduced number of defects per unit from .49 in January to .04 in December.
- Reduced number of defects per unit from .85 in January to .22 in December.
- Improved System yield from Q1 to Q2 average 63 percent to Q3 to Q4 average 78 percent.
- Failed P.C. Boards are returned in 24 hours.
- Hardware installation yield last four months at 100 percent.
- Hardware installation yield last three months out of last four at 100 percent.

COST REDUCTIONS

- Operations Finance nonvalue-added reduction 40 hours/week.
- S.P.C. and Ship to Stock decreased indirect requirements 200 hours/week.
- Material Handler put into direct 40 hours/week.
- System transactions reduced in Haverhill by 62 percent.
- Consolidated kitting and shipping functions reduced 120 hours/week.
- Labor content decreased standard 5.6 hours on certain product and 11 hours on another.
- Outside storage cost reduction.

INVENTORY TURN IMPROVEMENTS

- Laminate from 3 to 9.9 turns.
- P.C. Fab W.I.P from 14 to 28 turns.
- Haverhill from 4.6 to 6.1 turns.
- Haverhill total from 3.5 to 4.6 turns.

AVERAGE CYCLE TIME REDUCTIONS

■ P.C. Blanks	from 12.4 to 7.4 days
■ P.C. Boards	from 12 to 3.0 days
■ Options	from 5 to 3.0 days
■ Systems	from 7 to 3.8 days
■ System A	from 12 weeks to 8 weeks
■ System B	from 35 to 12 days
■ System C	from 40 to 12 days
■ System D	from 20 to 10 days

LOT SIZE REDUCTIONS

- P.C. Blank reduced 9 units from 36.
- Metal Fab lot sizes reduced from 48 days to 28 days.
- P.C. Boards reduced to 5 units from average of 50.
- Systems/Options reduced to one week maximum.
- Keyboards reduced to daily requirements.

SETUP REDUCTIONS

- P.C. Fabrication Drilling 50 percent reduction.
- P.C. Fabrication Solder Masking 25 percent reduction.
- N.C. Punch 75 percent to 80 percent reduction.
- 9600 frame weldments 100 percent elimination.
- Machine Shop identified 57 percent reduction.
- Paint Shop piece part change 52 percent reduction.
- Paint Shop color change 70 percent reduction.
- Paint Shop metal to plastic 73 percent reduction.
- P.C. Assembly Automatic Insertions 50 percent reduction.

"It is my opinion that the greatest contribution remains unmeasured. That contribution is the active participation and substantial development of increasing numbers of our employees. It is the future contribution from developed employee talents that will ensure our long-term success."

Dave Knight
JIT Implementation Manager

Figure 15-9. Improvements at Valleylab, Inc., resulting from JIT.

VALLEYLAB, INC.

MAJOR RESULTS

GENERAL/OVERALL PROGRAM

- Increased/Improved Communication
- Increased Employee Involvement
- Improved Employee Attitude/Commitment
- Cultural/Philosophical Exchange Evident

MATERIALS

- Raw Material Certification Program
- Delivery Reliability Increased (71 percent)
- Delivery Frequency Increased (100 percent)
- Acceptance Rate Improved (15 percent)

S.P. HARDWARE

- Cycle Time Reduced 83 percent.
  63 days to 10.4 days.
- Process Flow Mileage Reduced 55 percent.
  2,045 Feet to 925 Feet.
- WIP Inventory Reduced 28 percent.
- Capacity Increased 25 percent.
- Schedule Achievement Increased 24 percent.
  75 percent to 93 percent.
- Quality Improvement
  Subassembly Defects Reduced 31 percent.
  4.2 percent to 2.9 percent.

  Finished Product defects Reduced 51 percent.
  13.3 percent to 6.5 percent.

  Manufacturing Scrap Reduced 86 percent.

S.P. DISPOSABLES

- Quality Improved.
  Defects per million reduced 61 percent.
  Scrap and Rework Reduced 72 percent.
- WIP Inventory Reduced 33 percent.
- Capacity Increased.
  Final Assembly 12.5 percent.
  Mix Model Capability Handswitching Pencils 67 percent.
  3/Day to 5/Day
- Schedule Achievement Increased 15 percent.
  85 percent to 98 percent.
- Manufacturing Process Improvements:
  Process Flow Mileage Reduced.
  Cycle Time Reduced.
  Efficiencies (hours/unit) Improved.
  Line Balance Achieved.
  Machine Downtime Reduced.

Exhibit 15-1. Valleylab's pilot project plan.

VALLEYLAB

JUST-IN-TIME

PILOT PROJECT PLAN

SEPTEMBER 8, 1987

JUST-IN-TIME PROJECT

MISSION STATEMENT

Recognizing that its future growth and profitability depends significantly upon the effectiveness of its manufacturing operations, Valleylab is committed to certain key strategic plans in manufacturing. One such strategy is to develop knowledge, experience, methodologies, and business philosophies conducive to the implementation of Just-In-Time materials and manufacturing management throughout the company. The mission of the Just-In-Time Project Teams is to lead Valleylab to the accomplishment of this goal by directing educational efforts, by working to change management attitudes and business practices, by promoting greater worker involvement, and by demonstrating the benefits of Just-In-Time through successful implementation of selected pilot projects.

PROJECT SCOPE

The total project will be composed of two separate pilot projects conducted in the Surgical Products Division. One pilot will be in the disposables area and one in the hardware area.

Disposables

The projects included are the E2515, E2516, and E2502B handswitching pencils, with particular emphasis on the E2515 product line. The manufacturing processes encompassed in the project will be the final assembly and packaging operations. One E2515 assembly and one mixed model packaging line on each shift will be involved in the pilot project. The term of the project will be six months beginning August 1, 1987, through February 1, 1988.

Hardware

The project includes all regular production electro-surgery generators. This will include all Surgistat, 2L, Force 1B, Force 2, Force 100, and Force 4A and 4B generators. The manufacturing processes will not include printed circuit board or wire cable assembly operations performed outside the Surgical Products Hardware Department. All assembly, test, quality control, and packaging operations performed in the Surgical Products Hardware Department will be in the project scope. The project term will be nine months beginning August 1, 1987, through May 1, 1988.

JUST-IN-TIME PROJECT MISSION STATEMENT

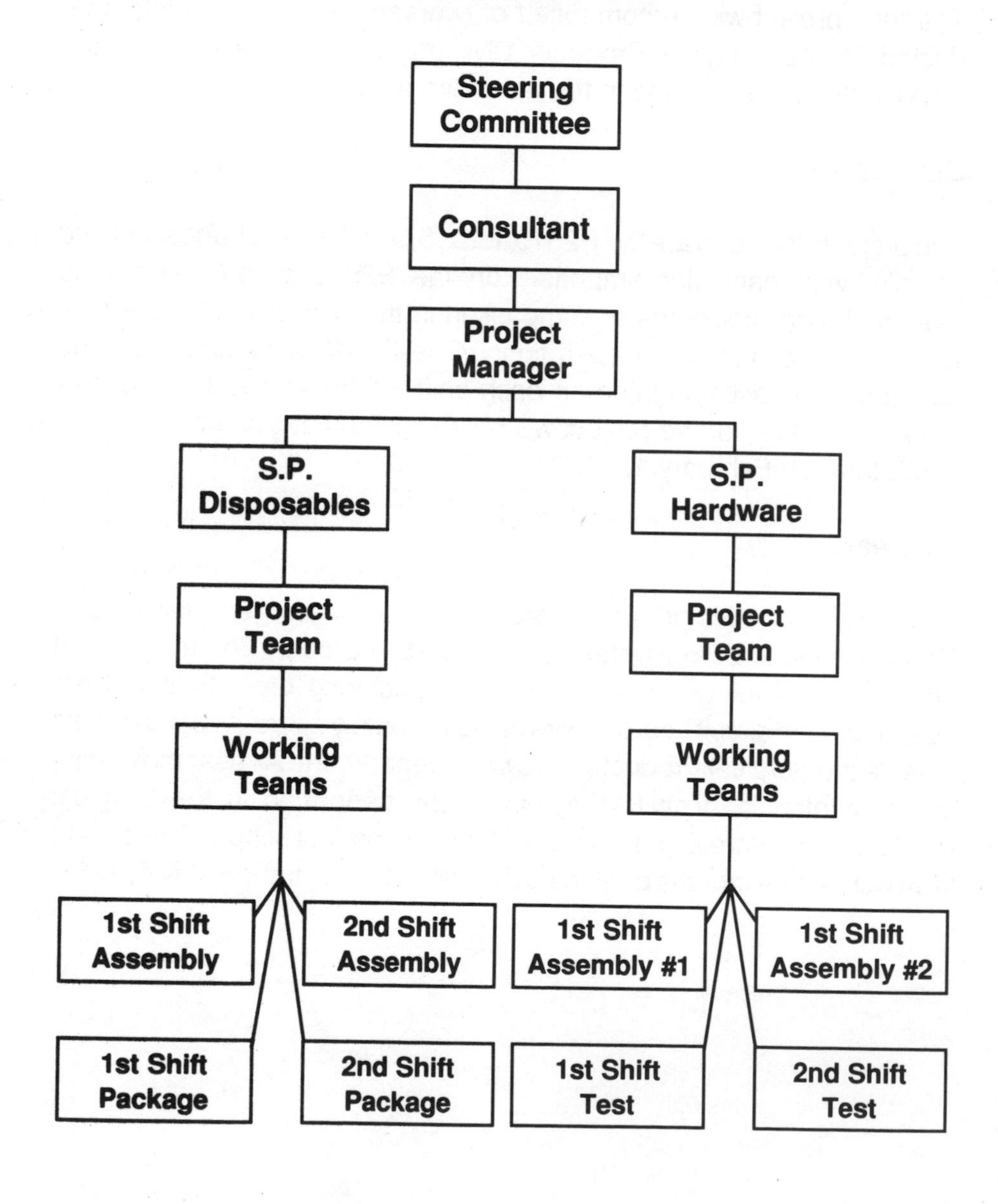

PROJECT GOALS

S. P. DISPOSABLES

Goal	Target
• Employee Involvement and Commitment	Increase
• Purchased Parts Procurement—"Drumbeat"	
Delivery Reliability	Increase to 95 percent
Delivery Frequency	Increase
Delivery Volume	Reduce
Acceptance Rate	Increase to 100 percent
Ship to Floor Program	Develop Guidelines
• Product Scheduling—"Drumbeat"	
E2515 Assembly Line	Achieve on 90 percent of days
F/F/S Package Line	at 95 percent of volume
Daily Mixed Model Mode	Achieve on 90 percent of days
• Quality Improvement	
Manufacturing Rework $s	Reduce 30 percent
Manufacturing Scrap $s	Reduce 30 percent
E2515 Assembly Reject Rate	Reduce 30 percent
F/F/S Package Reject Rate	Reduce 30 percent
Q.A. Sampling Defect Rate	Reduce 30 percent
• Work-In-Process Inventory $s	Reduce 33 percent
• Manufacturing Operations	
Efficiencies (hrs/unit)	Reduce 7 percent
Process Flow (mileage)	Reduce 50 percent

Goal	Target
Floor Space	Reduce 10 percent
Line Balance (Assembly Line)	Achieve balance
	Eliminate in-process queues
	Reduce cycle time 40 percent

- Documentation Review

Defining Document Review & Revise	Complete
F/F/S Packaging Specification Review & Revise	Complete
F2515 Assembly Specification Review & Revise	Complete
E2515H Assembly Specification Review & Revise	Complete
E2515–NSB Assembly Specification Review & Revise	Complete
E2516 Assembly Specification Review & Revise	Complete

S. P. HARDWARE

Goal	Target
• Employee Involvement and Commitment	Increase
• Purchased Parts Procurement—"Drumbeat"	
Delivery Reliability	Improve to 95 percent
Delivery Frequency	Increase
Delivery Volume	Reduce
Acceptance Rate	Improve to 100 percent
Ship to Floor Program	Develop guidelines
• Product Scheduling—"Drumbeat"	
Percent of Days Drumbeat Achieved	Increase 67 percent
Percent of Drumbeat Volume Achieved Daily	Increase 15 percent
• Quality Improvement	
Manufacturing Rework $s	Reduce 20 percent
Manufacturing Scrap $s	Reduce 20 percent
Subassembly Reject Rate	Reduce 20 percent
Final Assembly Reject Rate	Reduce 20 percent
• Work-In-Process Inventory $s	Reduce 50 percent
• Manufacturing Operations	
Process Flow (mileage)	Reduce 50 percent
Cycle Time	Reduce 75 percent
Capacity	Increase 25 percent
Efficiencies (hrs/unit)	Reduce 15 percent

Exhibit 15-2. Later Valleylab pilot project plan.

VALLEYLAB

JUST-IN-TIME

PHASE III PROGRAM

OCTOBER 1988

JUST-IN-TIME
MISSION STATEMENT

Recognizing that its future growth and profitability depend significantly upon the effectiveness of its manufacturing operations, Valleylab is committed to certain key strategic plans in manufacturing. One such strategy is to develop knowledge, experience, methodologies, and business philosophies conducive to the use of Just-In-Time materials and manufacturing management throughout the company. The mission of the Just-In-Time Project Teams is to lead Valleylab to the accomplishment of this goal by directing educational efforts, by working to change management attitudes and business practices, by promoting greater worker involvement, and by demonstrating the benefits of Just-In-Time through successful implementation.

SPD PROJECT
MISSION STATEMENT

As the Valleylab Just-In-Time philosophy matures, it becomes necessary to develop a communication network encompassing all aspects of materials and manufacturing management. The mission of the SPD Project Team is to establish communication lines and develop projects supporting worker teams by actively encouraging participation within and between all levels and interests associated with the Valleylab manufacturing process.

PROJECT SCOPE

This is "Phase III" of the Just-In-Time program for the Surgical Products—Disposables Division of Valleylab. The products included in the project are the E2515, E2516, and E2502B handswitching pencils. The manufacturing processes encompassed in the project will be the subassembly, final assembly, and packaging operations. The term of the Phase III project will be six accounting periods beginning AP10 (October 1988) through AP3 (March 1989).

Just-In-Time Program Organization

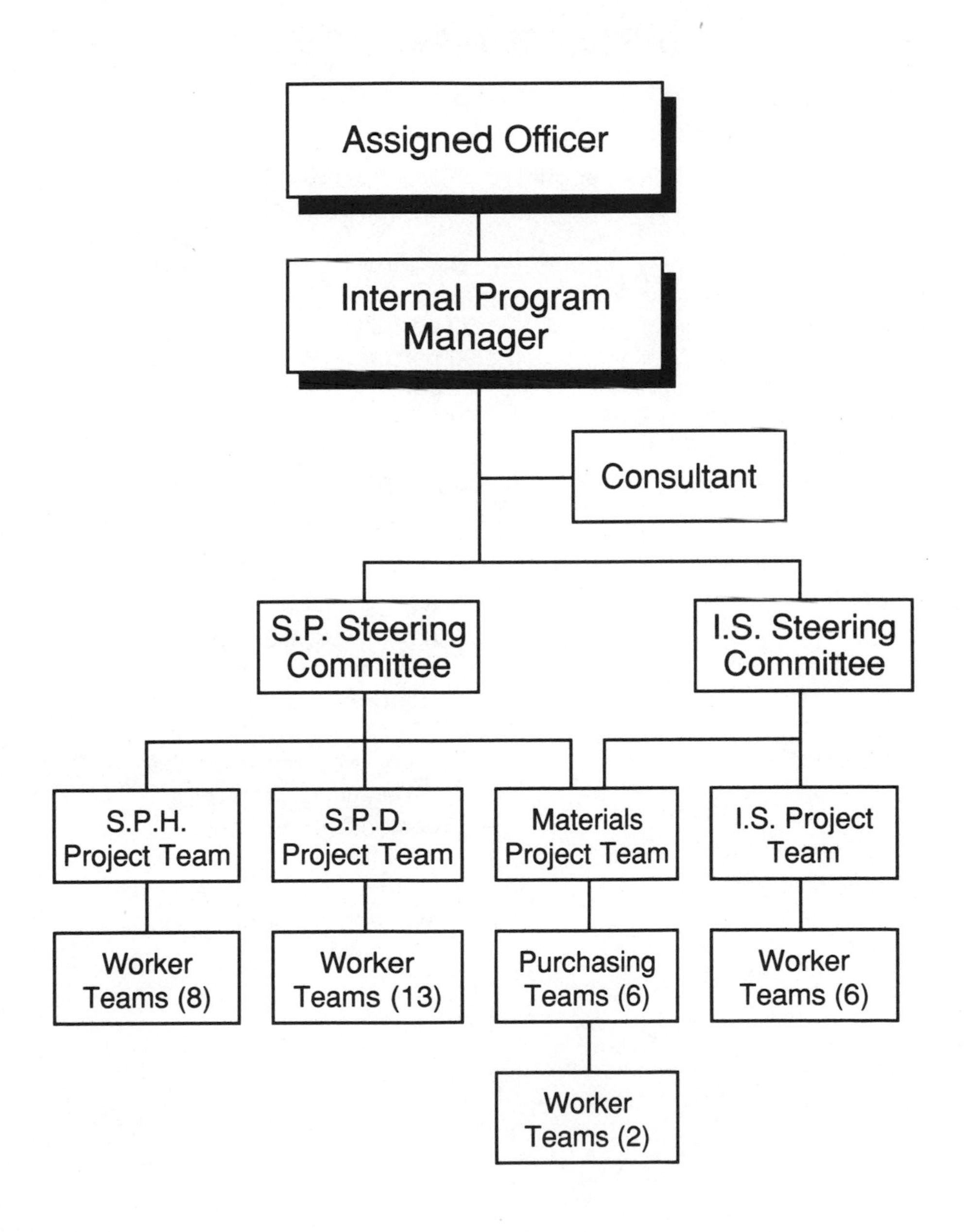

PHASE III GOALS

(AP10, 1988–AP3, 1989)

A. E2515 ASSEMBLY—First Shift

	Baseline	Goal
• Drumbeat	99 percent	100 percent
• Drumbeat Volume	90 percent	95 percent
• Manufacturing Defect Rate	1.2 percent	1.0 percent
• Q.C. Sampling Defect Rate (Goal is zero defects)	2 of 5AP	6 of 6AP
• Team Absentee Rate	2.1 percent	1.6 percent

	Goal
• Safety	• No lost time accidents • Monthly speaker, film, or demonstration concerning a safety item
• Job Enrichment	• Three programs (tour other departments, product demonstrations, etc.)
• Communication	• Write a newsletter every other month (goals, achievements, J.I.T. articles, etc.)

A. E2515 ASSEMBLY—Second Shift

I. <u>PRODUCT GOALS</u>

A. Drumbeat—to attain a weekly schedule using a flexible daily drumbeat.

B. Improve efficiency (hours per unit—1 percent).

II. <u>QUALITY GOALS</u>

A. Manufacturing Defect Percentage—Baseline is 0.35 percent. Goal is 0.3 percent.

B. "0" defect found by In-Process Quality Assurance inspectors.

III. <u>COMMUNICATION GOALS</u>

A. To develop a communication tool that will effectively present, maintain, and capture ideas, opportunities, problems, and solutions. The tool will be implemented within two months.

IV. <u>EDUCATION GOALS</u>

A. To train/retrain in the following areas:
1. Q.A. inspector training
2. Parts person training
3. Lead position training
4. Facilitator training
5. Other work-related job training

V. <u>SAFETY GOALS</u>

A. Have no lost time accidents

B. Have no OSHA recordable accidents

C. Receive safety-related training

VI. <u>MILESTONES</u>

On November 12, 1988, we will have gone one full year without a defect found by quality assurance inspectors.

A. E2515H/NSB Assembly—Second Shift

I. PRODUCTION GOALS
   A. Maintain a weekly volume with a flexible daily Drumbeat.

   B. Improve efficiencies (Hours per unit—3 percent).

II. QUALITY GOALS
   A. Manufacturing defect rates baseline 1.3 percent—Goal
      1. @ 2 months 1.2 percent
      2. @ 4 months 1.1 percent
      3. @ 6 months 1.0 percent

   B. Have "0" defects found by quality assurance inspectors.

   C. Have team go through quality assurance Inspector training.

III. TRAINING GOALS
   A. Increase knowledge of subassemblies
      1. Tour area
      2. Receive defect training

   B. Train/Retrain
      1. Lead position
      2. Parts position
      3. Trace/BV system—documentation

   C. Organize awareness sessions with various departments to gain better understanding of Valleylab
      1. Customer service
      2. Quality complaints
      3. Materials
      4. Engineering

IV. COMMUNICATION GOAL
   A. Implement an information board along the lines of the fishbone chart suggested by Charlene Adair.

A. F/F/S (Packaging)—Second Shift

I. PRODUCT GOALS

A. Drumbeat—Drumbeat will be separated into a weekly schedule with daily drumbeat and a volume drumbeat for the AP tied to the weekly schedule.

1. Attain Daily Drumbeat.

B. Improve Efficiency (Hours per unit: maintain—13 percent).

II. QUALITY GOALS

A. Manufacturing defect rates—baseline is 4 percent—goal is:

1. @ 2 months 3.8 percent
2. @ 4 months 3.6 percent
3. @ 6 months 3.4 percent

B. Have "2" defects found by Q.A. inspectors; baseline has been "3".

C. Conduct a retraining meeting(s) to assure manufacturing defects are consistently rejected within our team.

III. TRAINING GOALS

A. Receive/conduct training in following areas.

1. Duties of operators
2. In-Process Q. A. Inspection

IV. SAFETY GOALS

A. Have no accidents or injuries within the team.

B. Receive training in First Aid and CPR.

C. Set up a lifting technique training class before four months.

V. <u>COMMUNICATION GOAL</u>

A. Develop a communication tool within one month to track, keep visual, and keep informed on ideas, problems, and solutions. Implement within two months.

B. Create better communication lines with assembly teams.

VI. <u>MACHINE-RELATED GOALS</u>

A. Develop and implement an information tool to track machine breakdowns, causes, downtime, and preventive measures.

B. Offer training to team in operator-based preventive maintenance.

A. E2515 Overmold Area—First Shift, Team A

1. Attain Drumbeat
   Weekly 90 percent, volume 100 percent

2. In-Process Defects
   Currently at 1.2 percent
   We want 1.0 percent

3. Q.A. Defects
   For a 4 week AP 2 defects
   For a 5 week AP 3 defects

4. Improve Efficiency (hours per unit 3 percent)

5. Job Enrichment
   Tour other departments or invite speakers from other areas three times in six A.P.'s.

6. Develop and use Fishbone chart to improve communication

7. Safety—no lost time accidents in six months

8. Have team members start charting daily numbers so that we know how successful we are

A. E2502B Assembly—Second Shift

PROJECT SCOPE

This Phase III JIT Implementation Plan has been designed and written by the 02B JIT Team.

This document, therefore, defines the evolving 02B commitment to renew team project goals, to create a comfortable and efficient work environment, and to implement and maintain communication lines with other teams.

PRODUCTION SCHEDULE GOALS

To achieve the drumbeat schedule on 100 percent of scheduled days at 100 percent total volume per AP.

QUALITY IMPROVEMENT GOALS

To establish an assembly defect rate of .2 percent from a baseline rate of .5 percent.

To eliminate in-process retentions.

To eliminate final audit retentions.

To eliminate QA sampling defects.

MANUFACTURING OPERATIONS GOALS

To reduce the hours per unit 11 percent.

To develop a new product flow kanban system.

ENVIRONMENTAL GOALS

Redesign the final assembly solder table.

Redesign the conveyor belt.

Add recessed bins for blades and rockers in the existing base of the hot stamp.

Modify tail cap fixture.

Modify red taping fixture to accommodate six units instead of five.

Obtain adequate seating.

COMMUNICATION GOALS

To continually develop and maintain constructive dialogue within the 02B team.

To open and maintain dialogue with product engineering.

To open and maintain dialogue with Q.A.

To open and maintain dialogue with our vendors.

To open and maintain dialogue with Valleylab management.

S. P. HARDWARE JUST-IN-TIME GOALS

PRODUCTION SCHEDULE—"DRUMBEAT"

Daily drumbeat will be set by a schedule that is planned on an AP basis. Due to differing labor rates on the various products, this drumbeat will need to flex daily according to the particular product mix scheduled for the specific AP. Depending on the schedule, drumbeat will vary per day. This schedule will be for the entire generator production process. This schedule will determine daily drumbeat for PCB, subassembly, final assembly, test, Q.C., and packaging operations.

One goal of the program will be to achieve schedule on 80 percent of the total days each AP. Second, the goal will be to achieve 90 percent of the total volume each AP, without giving credit for any units produced in excess of the daily drumbeat schedule.

QUALITY IMPROVEMENT	REDUCTION %
• Defect Rates	
PCB:	
Subassembly	20%
Finished Product	50%
Accessories	50%
• Manufacturing Scrap	10%
• Manufacturing Rework	10%

WORK IN PROGRESS	REDUCTION %
• WIP Dollars:	
Generators	5%
Accessories	17%
• Reduce Cycle Time	

PCB FUNCTIONAL TEST

Four printed circuit boards have been identified for implementation of Functional Test setups. This will be accomplished by the end of AP5 of this year.

SUBASSEMBLY FUNCTIONAL TEST

Due to the success we have seen from PCB functional testing, we have identified four types of electrical subassemblies that can be inexpensively tested and result in a positive gain. These four subassembly setups will be implemented in AP3, AP5, AP7, and AP9 of 1989.

Exhibit 15-3. Pilot project for indirect teams.

THREE PLANS FROM INDIRECT TEAMS EACH WITH MEMBERS FROM RECEIVING, INSPECTION, AND THE WAREHOUSE

JIT SETTERS GOALS

I. Personal Involvement/Attitude

A. Attitude Survey

1. Administer employee attitude survey at the beginning of JIT program implementation and again at three, six, and twelve months.

2. Evaluate results in terms of positive attitude changes on the part of all warehouse personnel.

Schedule

Start	**Date**	**Tabulation of Results**
Survey I	October 28, 1988	January 5, 1989
Survey II	April 3, 1989	April 13, 1989
Survey III	August 7, 1989	August 17, 1989
Survey IV	December 4, 1989	December 14, 1989

B. JIT "I Make A Difference" Badges

1. Design, order, and distribute badges to all warehouse staff.
   a. Design badges—October 14, 1988
   b. Order badges—November 4, 1988
   c. Distribute badges—December 12, 1988

2. Seek personal commitment on the part of each individual to the successful completion of all warehouse JIT projects.

Schedule
Seek commitment—December 12, 1988

C. Bulletin Board

1. Arrange for the purchase and installation of a JIT bulletin board in the warehouse area.

2. Develop an effective communications format for the bulletin board display.

Schedule
Research started—October 18, 1988
Complete target date—April 27, 1989

D. Potluck Lunches

1. Organize warehouse-wide monthly "Birthday" potluck lunches to encourage:
   a. Better communication between groups.
   b. Team spirit.
   c. A more cost-effective way to recognize individual birthdays.

Schedule	**Start Date**	**Completion Date**
Group discussion	November 4, 1988	
Informal survey	November 11, 1988	December 8, 1989
Present time and cost savings	January 9, 1989	January 12, 1989
Resurvey warehouse	January 12, 1989	January 26, 1989
Finalize project		February 23, 1989

II. Zero Discounts Lost for Period of One Week

Initiate a team effort to expedite **ten day vendor** discount receipts through the receiving system. (Potential cost savings annually could be as high as $20,000 based on 1988 data.)

Schedule

Initial group discussion—October 28, 1988
Project finalization approval target date—November 4, 1988
Kickoff of discounts lost—December 12, 1988

III. Foam/Foil Packaging

Investigate better method to package materials.

Schedule

Initial discussion—December 8, 1988
Completion target date—July 27, 1989

IV. Stockroom Copier

Investigate the purchase or lease of a second copier to be located in the stockroom.

Schedule

Initial discussion—December 8, 1988
Completion target date—December 21, 1989

V. Safety

A. Work with members of other JIT teams to keep the warehouse safe.

1. Observe safety shoe use requirements.

2. Discourage use of the Receiving Inspection lab and MRB cage areas as a "shortcut" to or through the warehouse.

3. Investigate need for safety "boots" on racking corners and/or for bolting racks to the floor.

Schedule
Initial group discussion—December 8, 1988
Completion target date—August 31, 1989

VI. Training/Personal Enhancement

A. Initiate training programs in the following areas:

1. Copier use

2. Computer skills

3. English usage/writing skills

Schedule
Initial discussion—December 1, 1988
Completion target date—May 26, 1989

B. Seek approval for purchase and distribution of dictionaries to all work areas throughout the warehouse.

Schedule

Initial discussion—December 1, 1988

Completion target date—March 30, 1989

VII. Housekeeping

A. Work to keep the Warehouse/Receiving Inspection lab areas neat and clean.

B. Practice good housekeeping throughout the workday.

Schedule

Ongoing continuously

ROSS'S REBELS GOALS

Goals

I. Address Inventory Levels

A. Select 2 part numbers from each area (ISD, ISH, SPD, and SPH) and increase or reduce inventory levels to conform to "Material Planning Ordering Policy."

1. Establish monitoring methods January 1989.

2. Determine and assign monitors January 18, 1989.

3. First monitoring period will begin January 1989 and last through AP3.

II. Reduce Recurring I.R.'s

A. Understand the problems:

1. Open discussion in JIT group (January 1989).

2. Interview appropriate personnel from Q.E., Engineering, Planning/ Purchasing. This will be ongoing beginning in January and continuing throughout the year.

3. Establish short-term pilot project "Hipshot" (March 1989).

4. Analyze results of pilot project after six months (September 1989).

5. Proceed with project accordingly.

III. Communication

A. Establish better communication with other departments.

1. Invite guests to JIT meeting.

2. Attend JIT meetings of other groups.

3. Send memos.

This will begin in January of 1989 and continue throughout the year.

JIT BRAINSTORMERS GOALS

<u>**Goal**</u>

I. Daily to Weekly Timecard

<u>**Objective**</u>

A. To get all indirect labor employees from daily to weekly timecards.
Target Date: December 31, 1989.

1. Warehouse people as pilot run.
Target Date: December 31, 1988
Completion Date: November 22, 1988

2. Next department to be QC Receiving.
Target Date: January 3, 1989
Completion Date: January 3, 1989

3. Add other departments in two-week intervals.
Target Date: December 31, 1989

B. Cost savings to follow upon completion of program.
Target Date: December 31, 1989

<u>**Goal**</u>

II. Packaging/Delivery Routing

<u>**Objective**</u>

A. Direct vendor delivery to production of styrofoam.
Target Date: February 16, 1989.

1. Pilot program.
Target Date: February 16, 1989.

2. Change scheduled quantities.
   Target Date: March 1, 1989.

B. Quality Acceptance

1. Certify vendors: Styro-molders and Colorado Container.
   Target Date: October 31, 1989

2. Source inspection of product.
   Target Date: December 31, 1989

C. Maintaining Traceability

1. Change S.O.P. for tagging packaging.
   Target Date: June 31, 1989

2. Change styro's multiple part numbers to one per product.
   Target Date: September 31, 1989

Goal

III. Safety

Objective

A. Work together for a safer warehouse.

1. Foot protection
   Target Date: March 1, 1989
   a. Nonwarehouse people
   b. Placards
      1. Foot protection beyond this point.
         Target Date: March 1, 1989
         Completion Date: Received January 13, 1989.

c. Memo to department managers of new foot protection policy.

2. No lost time accidents.
   Target Date: Ongoing
   a. Training films.
      Target Date: 1 each quarter
   b. 500 club
      Target Date: 1990

3. Training of personnel in OSHA safety standards.
   Target Date: Ongoing

<u>Goal</u>

IV. Training/Personal Enhancement

<u>Objective</u>

A. Interaction with other departments.
   Target Date: Ongoing

1. Tours of other departments.
   Target Date: 1 each quarter.

2. Off-sight tours of other company warehousing.

3. Cross-training to other warehouse functions.
   Target Date: 1 person at least 1 hour per week.

B. Guest speakers/films and seminars.

1. New warehouse facilities.
   Target Date: February 21, 1989

<u>**Goal**</u>

V. Housekeeping

A. Work together to establish and maintain a clean warehouse.
Target Date: Ongoing

1. Practice good housekeeping fifteen minutes a day.

Index